IT LEADERSHIP

SEGVAN JOHNSON

From Visionary to Legend: IT Leadership That Shapes the Future

Copyright Statement

For permission requests, please contact the author at Segvan2026@gmail.com

ISBN: 979-8-9923955-0-1 (eBook)

ISBN: 979-8-9923955-1-8 (Paperback)

ISBN: 979-8-9923955-2-5 (Hardback)

Library of Congress Control Number: 2025900589

First edition January 2025

Chapter Overview

This book is structured into eleven parts, each addressing a critical dimension of IT leadership. From foundational vision to operational execution, from strategic foresight to organizational culture, each part builds on the last. Together, they equip you with the tools, frameworks, and real-world guidance necessary to lead with vision, impact, and lasting purpose.

Whether you're driving transformation in a legacy environment, scaling innovation in high-growth settings, or shaping the future of digital infrastructure, this book is designed to support both your strategic thinking and your execution discipline, chapter by chapter, part by part.

Copyright Statement 3

Chapter Overview 4

Dedicated to all the visionary leaders who inspire progress and innovation. 10

Acknowledgments 11

About the Author 12

About the Book 14

How to Use This Book 16

Lead the Future, Leave a Legacy 20

The Visionary Journey 21

Introduction: Visionary Leadership in IT 25

Part 1 32

The Visionary Foundation (Chapters 1–4) 32

Chapter 1 33
Visionary Leadership in IT 33
Chapter 2 51
Governance: The Foundation of Trust 51
Chapter 3 62
Mastering Digital Transformation 62
Chapter 4 81
Building a Secure Digital Fortress 81
Part 2 92
Leading Through Challenges (Chapters 5–7) 92
Chapter 5 93
Building Legendary Teams 93
Chapter 6 103
IT as a Catalyst for Business Growth 103
Chapter 7 113
Leadership During IT Crises 113
Part 3 121
Metrics and Sustainability (Chapters 8–9) 121
Chapter 8 122
Metrics That Matter 122
Chapter 9 133
Staying Ahead of the Curve 133
Part 4 143
Staying Ahead and Adapting (Chapters 10–13) 143

Chapter 10 144
The Legendary IT Mindset 144
Chapter 11 152
The Ethics of IT Leadership 152
Chapter 12 167
The Global Visionary IT Leader 167
Chapter 13 181
Driving Sustainability Through IT 181
Part 5 190
Engaging and Inspiring Others 190
(Chapters 14 - 15 - 16) 190
Chapter 14 191
IT's Role in Customer Experience Excellence 191
Chapter 15 201
Future Proofing IT Leadership 201
Chapter 16 211
IT Leadership as a Catalyst for Transformation 211
Part 6 220
Innovating for the Future (Chapters 17 - 18) 220
Chapter 17 221
Storytelling as a Leadership Tool in IT 221
Chapter 18 229
Cultivating a Culture of Continuous Innovation 229
Part 7 238

Legacy and Impact 238
(Chapters 19 - 21) 238
Chapter 19 239
Mentorship in IT Leadership 239
Chapter 20 251
AI Leadership: Harnessing the Future Today 251
Chapter 21 262
Leaving a Legacy of Transformative IT Leadership 262
Part 8 273
Leading Through Complexity 273
(Chapters 22–24) 273
Chapter 22: 275
Leadership During Technology Mergers and Acquisitions ... 275
Chapter 23: 284
Leveraging Data Driven Decision Making 284
Chapter 24: 293
Managing Remote and Hybrid IT Teams 293
Part 9 302
Transforming IT Leadership Through Innovation 302
(Chapters 25–26) 302
Chapter 25: 303
Leadership in Open-Source Technology Adoption 303
Chapter 26: 312
Harnessing Quantum Computing in IT Strategy 312

Part 10 319

Governance, Ethics, and Resilience in IT 319

(Chapters 27–30) 319

Chapter 27: 321

The Role of IT Leadership in Shaping Digital Democracies .. 321

Chapter 28: 328

The Role of IT in Disaster Recovery Planning 328

Chapter 29: The Art of Negotiation for IT Leaders 335

Chapter 30: 342

Legal Challenges and Compliance in IT Leadership 342

Part 11 351

Strategic IT Execution (Chapters 31–32) 351

Chapter 31: 352

Technologies and Tools Every IT Leader Must Adopt 352

Chapter 32: 361

Infrastructure Strategy and Vendor Consolidation 361

Conclusion 372

From Visionary to Legend 372

Appendices 376

- Appendix A 377
- Leadership Frameworks and Tools 377
- Appendix B 382
- Sustainability Practices in IT 382
- Resources for Further Reading 383

Your Action Plan 384
Checklist for IT Leadership Excellence 389
Your Leadership Reflection 394
The Last Word: Shaping Tomorrow with Vision 397
Thank You for Reading! 399

Dedicated to all the visionary leaders who inspire progress and innovation.

Dedicated to the visionary leaders who ignite change and drive innovation, and to those who aspire to lead with purpose, courage, and clarity.

To all who challenge the ordinary, empower others to rise, and illuminate the way toward a brighter, more inclusive future.

Your courage to question limits, your creativity to reimagine what's possible, and your determination to turn ideas into impact inspire change across industries, communities, and lives.

This book is a tribute to your vision, your resilience, and the legacy you are shaping, a legacy not only of success, but of significance.

May your leadership journey continue to inspire and guide others to step forward, lead boldly, and shape what's next.

Acknowledgments

Writing this book has been a journey of discovery, growth, and collaboration. It is a reflection of countless shared ideas, moments of inspiration, and the unwavering support of many along the way.

To everyone who has contributed to this endeavor, whether through thought-provoking conversations, insightful feedback, or simply offering encouragement, **THANK YOU**. Your perspectives and wisdom have shaped this work into something far greater than I could have achieved alone.

This book stands as a testament to the power of collaboration and shared purpose. It reflects the contributions of those who believe in the transformative potential of leadership, innovation, and progress. To all who inspire others to rise, to dream, and to create meaningful change, this is for you.

May the insights within these pages serve as a source of inspiration and guidance to leaders everywhere, and may they carry forward the spirit of collaboration and vision that brought this work to life.

About the Author

Segvan Johnson is a visionary IT strategist, digital transformation leader, governance expert, and infrastructure innovator who turns technology into a cornerstone of organizational success. With a blend of academic rigor and practical experience, he brings a forward-thinking perspective to enterprise architecture, automation, and strategic leadership. Segvan is currently completing his Doctor of Business Administration in Information Technology, combining scholarly depth with real-world insight to address the evolving challenges of today's digital landscape.

Throughout his career, Segvan has guided organizations through transformation, strengthened IT resilience, and aligned intelligent technologies, such as automation and AI with business strategy. His leadership philosophy centers on empowering people, nurturing creativity, and ensuring that technology serves a greater purpose, not just within businesses, but across society.

He is a strong advocate for open-source technologies, believing in their power to foster global collaboration, accelerate innovation, and deliver scalable, accessible solutions, especially in fast-evolving IT environments. Segvan also champions intelligent automation and inclusive innovation. He believes in technology's power to create opportunities, accelerate collaboration, and solve meaningful problems. A dedicated mentor, he champions knowledge-sharing as a foundation for long-term leadership and sustainable excellence.

In *From Visionary to Legend: IT Leadership That Shapes the Future*, Segvan shares practical insights, frameworks, and strategies drawn from years of guiding IT teams and executive leaders through high-stakes initiatives. Whether modernizing infrastructure, driving digital transformation, or building cross-functional alignment, his work provides actionable guidance to help others lead with clarity, purpose, and foresight.

Beyond his professional work, Segvan finds inspiration in emerging technologies, nature, and mentoring the next generation of IT leaders. His curiosity, sense of purpose, and reflective mindset shape everything he does, and serve as a reminder that leadership isn't just about solving problems; it's about creating meaningful, enduring possibilities.

This book reflects my journey, and I hope it guides yours.

About the Book

From Visionary to Legend: IT Leadership That Shapes the Future is an authoritative, inspiring, and practical guide crafted to empower IT leaders and aspiring leaders to redefine their roles in a rapidly transforming digital world. This book provides a roadmap for professionals who are ready to rise beyond traditional management and embrace their potential as innovators, change-makers, and visionaries shaping the future of their organizations and industries.

Every chapter dives deep into the challenges, opportunities, and responsibilities that come with modern IT leadership, including leading hybrid teams, managing vendor ecosystems, and navigating global complexity. Through actionable strategies, thought-provoking insights, and relatable examples, it equips readers with the tools to navigate a complex landscape of technology, people, and business. Whether you're leading a global digital transformation, building resilient teams, crafting a vendor strategy, or shaping a legacy of innovation, this book provides the clarity and confidence to succeed.

The foundation of leadership begins with vision. Chapter by chapter, readers are guided through the process of cultivating and refining a vision that aligns with organizational goals while driving sustainable, forward-thinking growth. From there, the focus shifts to execution, bridging governance, strategic alignment, risk management, and vendor strategy, to ensure that every initiative is built on a solid foundation of trust, accountability, and measurable impact.

Innovation and transformation are central themes. As the pace of technological change accelerates, the ability to adapt and lead boldly is critical. This book dives into key areas like mastering digital transformation, leveraging emerging technologies, embedding creativity into organizational culture, and measuring outcomes with meaningful metrics. Leaders are shown how to create environments where experimentation thrives, where risks are calculated rather than feared, and where breakthroughs become constant.

Recognizing that technology alone doesn't drive success, *From Visionary to Legend* emphasizes the human side of IT leadership. It highlights the importance of building legendary teams, teams that are diverse, collaborative, and capable of rising to any challenge in a hybrid or distributed world. Readers will learn how to nurture talent, inspire trust, cultivate growth, inclusion, and shared purpose, and leave a lasting impact through mentorship and empowerment.

Customer experience, sustainability, and ethical responsibility also take centre stage, offering guidance on how IT leaders can balance innovation with integrity. Whether implementing AI responsibly, improving cybersecurity, ensuring equitable access to technology, or reducing environmental impact, this book provides a blueprint for leadership that goes beyond immediate goals to contribute meaningfully to society and future generations.

At its core, this book is about creating a legacy. It challenges leaders to think bigger, not just about their current projects or roles but about the transformative impact they can have on their organizations, industries, and communities. The final chapters inspire readers to reflect on their journey, measure their progress, strategize for the next chapter, and take bold steps toward shaping a future defined by resilience, creativity, purpose, and measurable legacy.

From Visionary to Legend is for those who see leadership as more than a title; it's for those who see it as a calling. It's a guide for IT leaders and aspiring leaders who are ready to redefine what's possible, drive lasting change, and inspire others to do the same. With every strategy, story, and insight, this book serves as a trusted companion for those determined to make their mark and leave a legacy of excellence.

How to Use This Book

From Visionary to Legendary: A Strategic Guide for IT Leaders at Every Level

This book is not meant to be read once and shelved. It's designed as a working playbook, a field companion for IT leaders navigating real transformation, whether you're leading a team, scaling infrastructure, driving policy, or shaping long-term vision.

Where you start depends on where you stand.

For Senior IT Leaders (CIOs, CTOs, VPs, Heads of Technology)

If you own enterprise strategy, cross-functional alignment, or innovation at scale, start with chapters that equip you to lead forward under complexity.

Start with:

- **Chapter 1**: Visionary Leadership in IT
- **Chapter 4:** Execution Discipline
- **Chapter 14**: Stakeholder Management
- **Chapter 22:** Platform Thinking
- **Chapter 31**: Strategic Technology Tools
- **Chapter 32**: Infrastructure and Vendor Strategy

These chapters focus on direction-setting, large-scale orchestration, and turning foresight into action.

For Mid-Level Leaders and Rising Architects (Tech Leads, Infra Managers, Enterprise Architects)

If you translate ideas into action, lead teams through change, or shape architecture decisions, start with chapters that help you connect business goals to technical clarity.

Start with:

- **Chapter 5**: Turning Strategy into Operating Models
- **Chapter 9:** Zero Trust and Security Architecture
- **Chapter 10**: Observability and Monitoring
- **Chapter 19**: Talent Systems for Technologists
- **Chapter 30**: Legal and Compliance Guidance
- **Chapter 32**: Procurement and Infrastructure Scaling

These will sharpen your leadership toolkit, not just your tech stack.

For Tactical Operators and Aspiring Leaders (Senior Engineers, Ops Leads, Security Leads)

If you're executing projects, managing tooling, or looking to rise into strategic roles, use this book to start thinking like a leader.

Start with:

- **Chapter 2**: Decision Making Under Pressure
- **Chapter 7**: Automation, AI, and Service Management
- **Chapter 8**: Help Desk and End-User Experience
- **Chapter 13**: Metrics That Matter
- **Chapter 29**: Vendor Management and Cost Clarity

Then revisit Chapter 1 and 4 to start reframing your leadership posture.

The Leadership Flywheel

Use this book dynamically:

- When **building vision** → Start at Chapter 1
- When **designing systems** → Read Chapters 5, 6, 22
- When **fixing delivery** → See Chapters 4, 10, 12
- When **managing complexity** → See Chapters 14, 25, 32
- When **developing others** → Use Chapters 19, 26

You don't need to read in order. You need to read with purpose.

Pro Tip: Turn Book → Action

- Use one chapter per week in a leadership meeting.
- Assign chapters as prompts for your team's growth plans.
- Turn frameworks into templates (or ask us, we'll help).
- Keep your vision, execution canvas, and strategy map alive alongside this guide.

This book is not a lecture. It's a leadership instrument.

Use it to author the future.

Lead the Future, Leave a Legacy

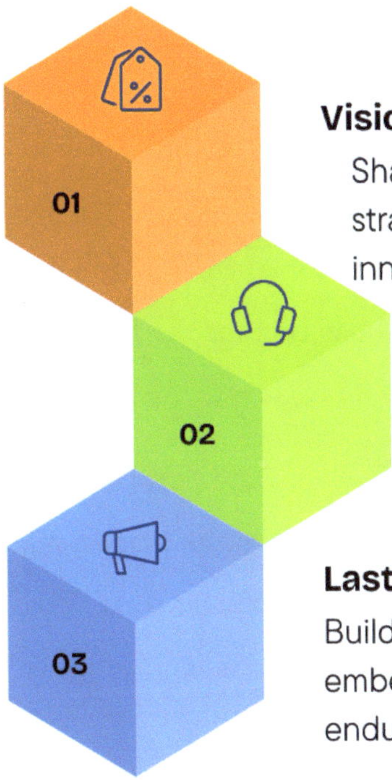

Vision & Purpose

Shape tomorrow with clarity. Align strategy with values to inspire innovation and progress.

Empowerment & Culture

Foster collaboration, curiosity, and integrity. Empower teams to create and lead with confidence.

Lasting Impact

Build systems, mentor leaders, and embed ethics to ensure your legacy endures beyond your role.

Lead the Future, Leave a Legacy

True leadership goes beyond managing the present, it shapes what comes next. As an IT leader, your influence extends far beyond infrastructure and systems; it touches people, culture, and industry evolution. The decisions you make today become the foundation of tomorrow's innovation, resilience, and progress.

To lead the future, you must act with clarity and purpose. Vision is not enough, it must be paired with strategic execution and a mindset that embraces change. Inspire your teams to explore what's possible, empower them to solve challenges creatively, and cultivate a culture that values curiosity, agility, and integrity.

Legacy isn't built on projects alone. It's reflected in the leaders' you mentor, the inclusive cultures you shape, and the systems that endure long after your role evolves. It's the ability to embed continuous learning, shared ownership, and ethical responsibility into the DNA of your organization.

Start now. Lead decisively. And build a legacy that doesn't just mark your time in leadership but defines it.

The Visionary Journey

Imagine a bustling healthcare organization at the height of a public health crisis. Patients flood in, medical staff are stretched to their limits, and the organization's reliance on technology is at an all-time high. From patient records to scheduling systems, every facet of care hinges on IT infrastructure. Then, seemingly out of nowhere, a ransomware attack cripples the hospital's digital ecosystem. Critical data is encrypted, and life-saving systems are paralyzed. Panic spreads through the organization like wildfire.

At the heart of this chaos stands David, the CIO. As alarms blare and the executive team demands answers, David doesn't retreat or succumb to panic. Instead, he takes command, his years of preparation and strategic foresight coming into focus.

He immediately activates the crisis response plan framework he meticulously built months prior with input from cross-functional teams. With a calm, steady voice, David rallies his IT staff, security experts, and operational leaders. Priorities are established: restore patient-critical systems, protect unaffected data, and communicate transparently with staff and external stakeholders. His clear direction and unflappable demeanor inspire confidence in everyone around him.

But David's leadership goes beyond containing the crisis. He sees the situation as an inflection point, a chance to reshape the organization's approach to IT. While teams work around the clock to regain control of systems, David collaborates with executives to secure investment for future-ready technologies, such as advanced threat detection and AI-driven incident response. He advocates for robust training programs, ensuring that employees can recognize and thwart potential attacks.

In the weeks that follow, not only does the organization recover, but it emerges stronger. The new safeguards implemented under David's guidance set a gold standard for the healthcare industry. Patients regain trust, knowing their data and care are secure. Staff operate with renewed confidence, bolstered by the systems David's team designed to withstand future challenges.

This story is not just about crisis management; it's about leadership that transcends the moment. David's actions exemplify the mindset and vision required of today's IT leaders. He didn't just react to the immediate problem, he saw the crisis as an opportunity to innovate, to set new benchmarks, and to ensure that the organization would thrive in the long term.

Leadership Rooted in Vision

The journey of an IT leader is one of purpose, resilience, and transformation. It's about moving beyond technical expertise to become a catalyst for change, a visionary who inspires teams, drives innovation, and navigates uncharted territories with confidence.

In a world where technology shapes every facet of life, IT leaders are at the forefront of solving complex challenges and creating opportunities that redefine industries. Visionary leadership begins with a compelling vision, a clear picture of the future that aligns with organizational goals and inspires action. It evolves through empowering teams, fostering inclusivity, and championing a culture of continuous learning.

This book is your guide to mastering that journey. It explores how to craft a lasting legacy by aligning strategy with purpose, embedding sustainability and ethics into decision-making, and mentoring the next generation of leaders. Whether you are a seasoned executive or an emerging leader, your journey is unique, and your potential to create impact is limitless. The path to becoming a visionary leader begins today are you ready to take the first step?

Visionary Leadership in IT

GUIDING ORGANIZATIONS THROUGH INNOVATION, CHANGE, AND PURPOSE

Introduction: Visionary Leadership in IT

Visionary leadership in IT is the compass guiding organizations through a landscape of constant disruption, innovation, and opportunity. IT leaders are no longer just technical managers; they are architects of the future, setting a course for progress, resilience, and transformation. This journey is not about managing systems but inspiring people, creating strategies, and turning challenges into opportunities for growth.

To lead in IT today, you must adopt a mindset that balances technical expertise with emotional intelligence, strategic foresight with adaptability, and innovation with a relentless focus on outcomes. This introduction lays the groundwork for understanding how visionary IT leadership can be the difference between mediocrity and extraordinary impact.

The Evolution of IT Leadership

The role of IT has evolved dramatically. What was once a support function ensuring systems were operational has become a strategic cornerstone of modern business. IT leaders are at the nexus of technology, business strategy, and human dynamics. Your decisions no longer just keep the lights on they shape the trajectory of your organization.

This transformation demands a shift in how you approach leadership. It requires you to step into a role that is as much about vision and communication as it is about coding and infrastructure. You are tasked with anticipating future trends, navigating crises with composure, and empowering your teams to innovate fearlessly.

Why Vision Matters in IT Leadership

At its core, visionary leadership is about setting a clear, compelling direction for your organization and rallying your team to pursue it with focus and determination. Vision is the foundation upon which every successful IT initiative is built. Without it, even the most innovative projects can falter, and even the most skilled teams can lose their way. A strong vision is more than an abstract ideal, it's a tangible roadmap that aligns technology initiatives with business objectives and resonates with stakeholders.

A powerful vision:

- ✓ Defines success in clear, actionable terms.
- ✓ Inspires trust, motivation, and commitment.
- ✓ Provides a framework for decision-making, even in times of uncertainty.

Your vision should articulate not just where your organization is going, but why it matters. It must resonate with your team, your stakeholders, and your customers. Aligning technology with broader business objectives, you create a roadmap that ensures every effort contributes to meaningful outcomes.

The Traits of a Visionary IT Leader

To be a visionary leader, you must cultivate traits that go beyond technical expertise. You need courage to take risks, clarity to make tough decisions, and empathy to inspire your team. These traits are not innate; they are developed through intentional effort and self-reflection.

1. **Adaptability:** The technology landscape changes constantly. As a leader, you must anticipate shifts, embrace change, and guide your organization through transitions with confidence.
2. **Empathy:** Technology is ultimately about people. Empathizing with your team, stakeholders, and customers enables you to create solutions that are not just functional but impactful.
3. **Strategic Foresight:** Visionary leaders think ahead. They identify emerging trends and opportunities, ensuring their organizations are always one step ahead of competitors.
4. **Resilience:** Crises are inevitable in IT, from cybersecurity breaches to system outages. Your ability to stay calm, assess the situation, and lead decisively will define your leadership.
5. **Collaboration:** No leader can succeed alone. Building strong relationships across teams, departments, and industries multiplies your impact and fosters innovation.

Building the Bridge Between Vision and Action

Having a vision is only the beginning. The true test of leadership lies in translating that vision into action. This requires a strategic approach, clear communication, and a focus on execution.

Start by breaking your vision into actionable goals. Each initiative should have measurable outcomes, clear timelines, and defined roles. Regularly communicate progress to your team and stakeholders, celebrating milestones and addressing challenges transparently.

Remember that execution is not a one-time effort, it is an ongoing process. You must be agile, ready to pivot as circumstances change, while keeping your overarching vision intact.

Empowering Your Team

Your vision is only as strong as the team that brings it to life. To lead effectively, you must create an environment where your team feels empowered, valued, and motivated. This means fostering a culture of trust, encouraging collaboration, and providing opportunities for growth.

Support your team with the tools, training, and autonomy they need to succeed. Recognize their contributions and invest in their professional development. Building a team that shares your vision, you create a force that drives innovation and delivers results.

Inspiring Confidence and Building Trust

Trust is the currency of leadership. Without it, even the most compelling vision will fail to gain traction. Building trust requires transparency, integrity, and consistent communication.

Be honest about challenges and decisions. Share the reasoning behind your strategies, and invite feedback from your team. Involving others in the process, you foster a sense of ownership and commitment that strengthens your vision.

The Role of Innovation in Visionary Leadership

Innovation is the heartbeat of IT leadership. It is not just about adopting new technologies it is about creating value through bold ideas and creative problem-solving. To be an innovator, you must foster a culture where experimentation is encouraged, and failure is seen as a steppingstone to success.

Encourage your team to think beyond immediate solutions. Challenge them to consider how their work contributes to long-term goals. Embedding innovation into your leadership approach, you ensure your organization remains competitive and resilient.

Balancing Short-Term Wins with Long-Term Goals

Visionary leadership requires balancing immediate needs with future aspirations. While it is important to deliver short-term results, these should always align with your larger vision.

Set priorities that address pressing challenges while paving the way for sustainable growth. Use short-term wins to build momentum and demonstrate the value of your vision. Each success should reinforce your team's commitment to long-term objectives.

Overcoming Resistance to Change

Change is a constant in IT, but it often meets resistance. As a leader, it is your role to guide your organization through transitions with empathy and clarity.

Address concerns proactively by explaining the benefits of change and involving your team in the process. Provide training and resources to ease transitions, and celebrate successes along the way. Leading with empathy, you turn resistance into collaboration.

Your Role in Shaping the Future

As an IT leader, you are not just managing today's challenges, you are shaping tomorrow's possibilities. Your vision has the power to influence not just your organization but the broader industry and society as a whole.

Take time to reflect on the impact you want to make. **How will your leadership drive progress, empower people, and create value?** Staying true to your vision and leading with purpose, you can create a legacy that endures.

A Call to Action

Visionary leadership is not a destination, it is a journey of constant growth, learning, and impact. Embrace the challenges, celebrate the successes, and remain steadfast in your commitment to making a difference.

Your role as an IT leader is more critical than ever. The decisions you make, the vision you champion, and the teams you inspire will define the future. Now is the time to step boldly into your role as a visionary leader and guide your organization toward extraordinary success.

The journey begins now. Are you ready to lead?

Part 1

The Visionary Foundation (Chapters 1–4)

Part 1 sets the stage for ***transformative IT leadership*** by establishing a strong foundation rooted in vision, accountability, and resilience.

Chapter 1 explores the ***principles and traits*** that define visionary leadership, emphasizing the importance of foresight, strategic thinking, and the ability to inspire others. It highlights how a compelling vision guides organizations through complex challenges and paves the way for innovation.

In Chapter 2, the focus shifts to ***governance***, offering strategies to build trust and accountability through transparent practices, compliance frameworks, and ethical oversight. Leaders are guided to foster a culture of integrity that underpins all decision-making.

Chapter 3 provides a roadmap for mastering digital ***transformation***, helping leaders navigate organizational change, integrate new technologies, and achieve sustainable success.

Finally, Chapter 4 equips leaders to secure their ***digital enterprises***, presenting robust strategies to safeguard assets, data, and operations in a rapidly evolving threat landscape.

Together, these chapters lay a comprehensive foundation for IT leaders to drive meaningful progress and inspire confidence in their teams and stakeholders.

Chapter 1

Visionary Leadership in IT

Visionary leadership drives every major IT transformation. It distinguishes those who simply manage systems from those who define what comes next. In the fast-moving world of information technology, effective leadership means navigating uncertainty, setting direction, and mobilizing teams toward outcomes few can yet imagine. It means shaping change rather than reacting to it. Visionary leadership does not begin with technology, it begins with decisions, and the ability to shape what's next before others see it coming. These leaders do more than guide organizations; they ignite momentum that transforms how technology and business grow together. In a time when progress moves faster than ever, the ability to align action with emerging opportunity becomes the most decisive advantage.

At this moment, as you begin reading, pause and note: visionary leadership is measurable. Treat it as a rhythm, not a trait, a weekly cadence of scanning, deciding, and signaling. Build a system around it, the same way you would any operational process.

The foundation of visionary leadership begins with a single question: **What is possible?** The most effective IT leaders do not wait for change to arrive, they create the conditions for it. They look beyond immediate obstacles and imagine how tomorrow could operate: high-performing teams empowered by meaningful tools, technology woven into strategy, and digital experiences that redefine customer expectations. Possibility-thinking requires more than optimism, it demands strategic curiosity and readiness to execute toward what could exist, even when no one is asking for it yet. Turning this into action means building mechanisms to surface possibilities weekly, across customer insight reviews, trend digests, and peer challenge sessions, then scoring them by urgency and impact. These leaders do more than guide organizations; they ignite momentum that transforms how technology and business grow together. In a time when progress moves faster than ever, possibility must be met with preparation, or opportunity becomes missed potential.

Treat the "**what is possible**" mindset as a weekly practice, not a philosophical slogan. Create a living backlog of horizon ideas sourced from cross-functional patterns, not product Wishlist, but structural friction maps.

Execution Guidance: use a 60-minute **Horizon Review** every Friday, one segment for new possibilities, one for threats, one for adjacent ideas seen in other industries. Capture no more than five items; the constraint forces clarity. In public sector or finance, assign one owner per horizon signal to test feasibility within 10 business days.

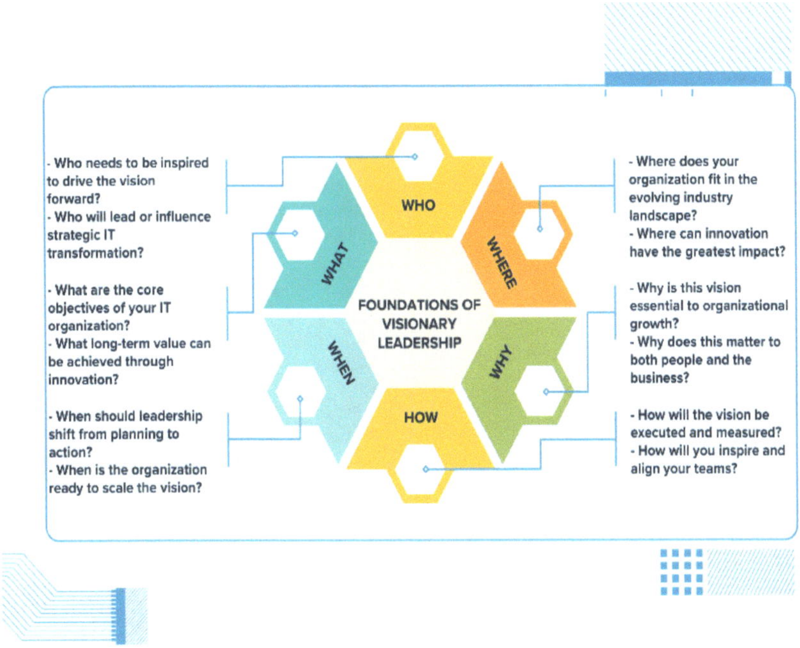

The Foundations of Visionary Leadership

Visionary leadership in IT is not born out of technical expertise alone. While mastery of technology forms a foundation, true leadership transcends technical skills, requiring an ability to connect innovation with purpose and strategy. A visionary leader operates at the intersection of technology and business, bridging the two to drive outcomes that are both impactful and sustainable.

The IT sector is inherently fluid, with emerging technologies like artificial intelligence, blockchain, and quantum computing continually altering the landscape. Yet, these advancements do not transform organizations on their own. What changes an enterprise is not the introduction of a new platform or tool, but how leaders orchestrate its application to drive measurable change tied to business performance. It is the leaders who anticipate the potential applications of these innovations, adapt strategies accordingly, and align their teams with a shared vision who bring true transformation.

Build a friction-to-fit map across your operational domains, what's slow, error-prone, or repetitive? Then overlay 3–5 technologies from your radar. That's not trend-watching; that's application design.

Visionary leadership is characterized by clarity of thought combined with an ambitious outlook. This clarity involves asking critical questions: **How are current trends reshaping industries? Where does my organization fit within this evolving picture? What long-term value can be achieved by leveraging these changes?**

Add one strategic foresight session per quarter where these three questions are asked across real incidents, market signals, and competitor shifts. Document assumptions. Make strategic bets.

Real-world application: in healthcare, use this session to map telehealth readiness vs. patient experience metrics; in manufacturing, test automation adoption against production variability; in financial services, review emerging regtech platforms that reduce audit cost or compliance friction.

The leader's role is to frame the business model impact of trends before they're fully visible, linking them to process improvements, cost transformation, experience redesign, or future revenue architecture. Addressing these questions enables leaders to transition from reactive managers to proactive visionaries.

To operationalize this foundation, start by scanning three lenses regularly:

1. Market inflection points in your industry.
2. Friction zones inside core processes.
3. Repeatable user pain points.

Then ask: which strategic technologies can address these with scale, speed, or new value? This creates a strategic trigger list, not just a backlog of tech ideas, but a prioritization of what to test, where to build, and how to prove business fit.

Store your trigger list in a shared workspace with ownership assigned to each item, turn abstract watchlists into operational bets, complete with impact hypotheses and lightweight test cycles.

Playbook Action: Evaluate one trigger per quarter. Use a **3x3 model**: 3 tests (low-cost pilots), 3 metrics (impact, speed, adoption), 3 weeks (to decide next step). This discipline converts vision into traction.

VISIONARY IT LEADERSHIP

Turning Innovation into Impact

01 THE IT SECTOR IS INHERENTLY FLUID

Emerging technologies like artificial intelligence, blockchain, and quantum computing continually reshape the landscape. But these innovations alone do not create transformation.

02 IT IS THE LEADERS WHO ACT

By anticipating how to apply these technologies, aligning strategies, and guiding teams with a unified vision, they unlock real transformation.

03 VISION BEGINS WITH CLARITY

When leaders ask the right questions about industry trends, positioning, and long-term value, they shift from reactive managers to proactive visionaries.

Defining Your Vision: The Compass for Leadership

A compelling vision is not an abstract declaration; it is a cohesive strategy that guides every decision. This vision is not built in isolation but arises from a deep understanding of the organization's goals, industry dynamics, and the evolving technology landscape. Visionary IT leaders ground their strategies in practicality while aiming for transformational outcomes.

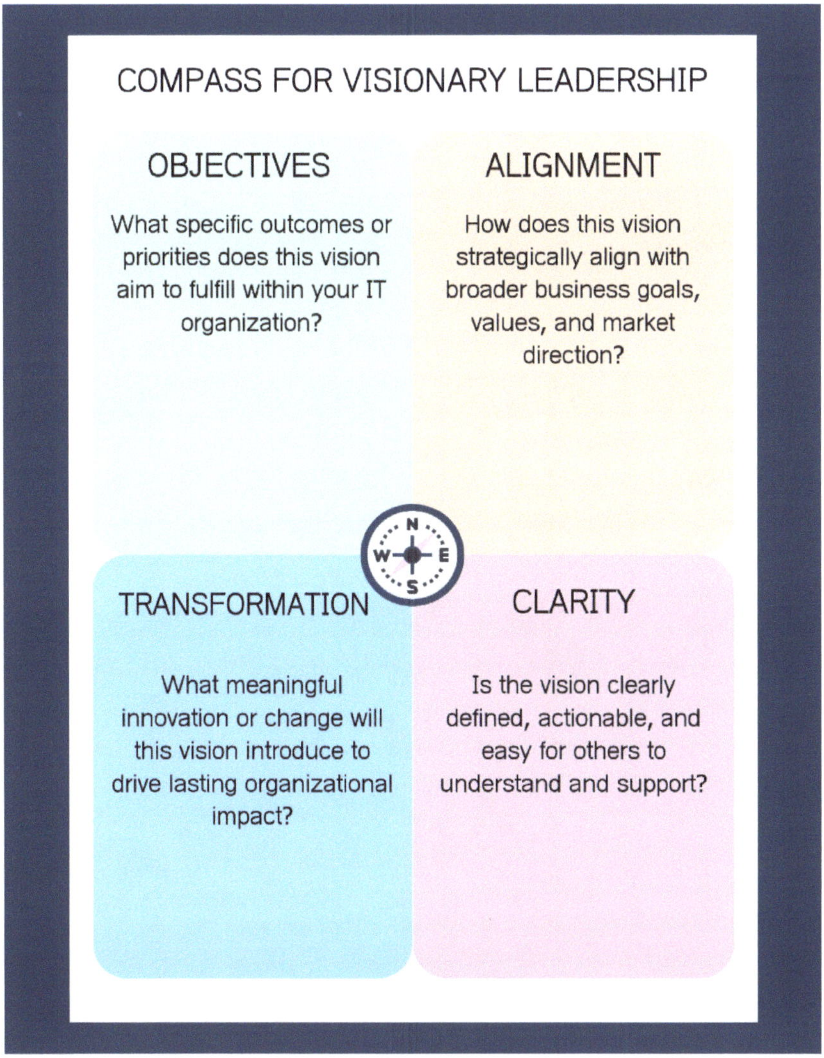

Crafting such a vision begins with reflection. What are the core objectives of your IT organization? Are you pursuing operational efficiency, creating competitive advantages, or fundamentally transforming customer experiences? Define outcomes before investments. Examples include reducing manual effort by 40% in finance workflows, launching multilingual support chat within 120 days, or achieving 90% uptime in self-service claim intake. Specificity matters. Always keep a rolling three-sentence version of your current vision active. Revise it quarterly based on new insight. Make it shareable, clear, and owned across teams.

Practical step: Translate each vision into one **execution headline** that could appear on a future performance report, e.g., Customer onboarding reduced from 14 days to 4, Claims decisioning automated with 96% accuracy. These become evidence of leadership foresight.

Equally important is alignment with broader organizational goals. IT leadership is not an isolated discipline; it operates as an enabler of business success. Visionary leaders clearly communicate how their technological initiatives support organizational growth, improve customer satisfaction, or bolster competitive positioning.

Tie every technology goal to a business-level statement: who it helps, what pain it relieves, what value it protects, or what capability it creates. If the vision cannot answer these four questions clearly, it's still premature.

Use a simple template across initiatives: "**This enables ___ to solve ___ so they can ___, which delivers ___.**" Vision gets traction when it's usable in everyday decisions.

Once defined, codify your vision into a simple three-part framework:

1. The strategic ambition (what future state you're building),
2. The technology pathway (what architecture or capability it requires),
3. The value thesis (how this impacts the business quantitatively and qualitatively).

Use this framework as the front page of your vision artifacts, executive memos, initiative briefs, or team readouts, so vision is always framed in outcomes.

Leadership habit: review this 3-part framework in every leadership meeting for 90 days, repetition hardwires clarity. When the framework becomes reflexive, your team starts anticipating the "why" behind every technology choice.

Building a Culture That Amplifies Vision

Even the most forward-thinking vision can falter without a culture that supports it. Organizational culture is the medium through which visionary ideas take root and grow. A leader's role, therefore, extends beyond crafting the vision to fostering an environment where it can thrive.

Implementation tip: start every quarter with a **culture calibration** check-in, what behaviors accelerated last quarter's outcomes, which ones slowed progress, and what norms need reinforcement? Culture drift is measurable if tracked deliberately.

This begins with authenticity. A leader's actions set the standard for the organization. **Are you demonstrating curiosity, collaboration, and courage in your day-to-day decisions?** Your team will reflect the qualities you embody. Furthermore, an inclusive approach ensures that every member of the team feels connected to the vision. **Transparency**, open communication, and active listening are essential in creating a culture where innovation flourishes.

Operationalize transparency by creating a **Vision Wall** where priorities, tradeoffs, and test results are visible and refreshed monthly. Culture grows in what people see, not what they're told.

Storytelling is another critical tool. Leaders must articulate their vision in ways that resonate deeply. Presenting a narrative that connects technological goals with real-world impact inspires teams to see themselves as contributors to something larger. Avoid technical monologues and instead translate strategy into user outcomes, experience upgrades, time savings, or ecosystem access.

Run **vision remix** exercises quarterly where teams rewrite the vision in their own words or pitch it back from the customer's lens. Alignment becomes owned, not enforced.

Establish cultural momentum by embedding rituals that keep the vision alive, weekly leadership syncs tied to vision progress, demo days where teams present working solutions, or even a **what's working** wall for success sharing. Culture doesn't need cheerleading, it needs scaffolding. Codify 3–5 rituals that reinforce direction, learning, and experimentation. Audit them yearly.

In industries like finance or manufacturing, supplement rituals with "shared metrics dashboards" visible to all contributors, transparency replaces hierarchy in reinforcing accountability.

Influence Over Authority: Leading with Conviction

In IT leadership, influence often outweighs authority. While authority enforces compliance, influence secures trust and commitment. Visionary leaders inspire their teams to buy into the vision not because they are required to, but because they genuinely believe in its purpose and feasibility.

Building influence begins with establishing credibility. Teams need to trust that their leader not only understands the challenges at hand but also has the foresight and resilience to navigate through them. This trust is built on **transparency, consistency, and integrity**. Publish early insights, admit unknowns, and surface learnings in real time. Make credibility visible. People trust what they see you wrestle with, not just what you present as finished.

To create credibility, show mastery of business impact, not just infrastructure design or budget oversight. Speak in outcomes and deliver in sprints that prove those outcomes early. Confidence follows visible traction.

Field move: in government or regulated sectors, begin **proof-first** habits. Every strategy brief must be anchored in one working pilot, no matter how small. Use that result as a credibility wedge, not theory, but evidence.

Empathy is equally important. Effective leaders invest time in understanding their teams, their aspirations, challenges, and motivations. Run regular **clarity syncs** with team leaders, not just about progress, but about purpose: **What's unclear in the strategy? What feels disconnected? What's not landing as intended?** Influence begins with perspective.

They connect the broader vision to individual goals, showing each team member how their contributions drive meaningful outcomes. This approach fosters a sense of personal investment that amplifies collective efforts.

Collaboration further strengthens influence. Leaders who engage their teams in shaping strategies, soliciting feedback, and incorporating diverse perspectives create a sense of ownership. This shared involvement transforms the vision from a top-down directive into a collective mission.

Make your strategy modifiable. Use open decision logs, participatory roadmap votes, or architecture discussions where engineers and designers shape direction, not just deliver it.

Leadership system tip: set a quarterly **strategy remix** challenge. Invite 2–3 rising team members to reframe a key part of the roadmap using current data, cross-functional feedback, and customer signals. Influence scales when ownership scales.

The Role of Strategic Foresight

In an industry characterized by rapid change, the ability to anticipate the future is indispensable. Visionary leaders do not simply adapt to trends; they leverage foresight to position their organizations ahead of the curve. This requires a deliberate approach to staying informed and analyzing the broader implications of technological advancements.

Staying informed involves more than consuming news or attending conferences. It requires analytical rigor identifying patterns across industries, evaluating emerging technologies, and understanding how they intersect with your organizational objectives. Create a lightweight internal **signals team** or rotating foresight group that meets monthly to share weak signals, draft future scenarios, and map responses. Don't chase hype, build interpretation muscles.

Use a consistent foresight method like:

1. Signals: weak but persistent shifts in tech, behavior, or regulation.
2. Scenarios: plausible futures based on combinations of signals.
3. Shifts: strategic moves you'd make if those futures proved true.

Scenario planning is an essential tool for foresight. Considering multiple futures, whether technological breakthroughs or market disruptions enables leaders to prepare adaptive strategies that ensure resilience.

Try building three future-state narratives per year, optimistic, conservative, disruptive, and use them in quarterly vision reviews. Foresight becomes relevant when tied to execution readiness.

To institutionalize foresight, create a quarterly **tech radar** document tailored to your organization. Not a generic trend list, but a living map of what's coming, what it means, and what needs testing. Turn that radar into a planning artifact. Make it mandatory for annual strategy and embedded in quarterly investment prioritization.

Action embed: in sectors like healthcare or manufacturing, map foresight not just to innovative teams but to operational leaders. Ask: **If this scenario happens, how does it impact scheduling, supply chains, service levels, or compliance?** Cross-functional foresight = strategic resilience.

Communication: Bridging Vision and Execution

A vision without effective communication risks being misunderstood or ignored. Visionary leaders must not only articulate their ideas clearly but also translate them into actionable plans that resonate with diverse audiences, from technical teams to executive stakeholders.

Tailoring communication is key. When addressing executives, emphasize strategic business impact, how initiatives support revenue, customer retention, or risk mitigation. For product teams, focus on outcomes and opportunity. For technical teams, detail the architecture and metrics that define success.

Don't send vision decks. Build living dashboards that show what's shipping, who it impacts, and what it unlocks. Let people follow the vision as a story unfolding, not a story told.

Never present an initiative without its:

- **Business case** (what changes),
- **User value** (who benefits),
- **Stakeholder map** (who's affected),
- **Strategic fit** (why now).

Consistency in communication reinforces momentum. Use multiple channels, weekly standups, monthly town halls, async briefs, and reinforce the same message with evolving context.

Use the **Rule of 7**: people need to hear a strategic priority seven different times and formats before it feels internalized. Repetition without variation creates clarity, not boredom.

Sector-specific tactic: in regulated industries, pair each vision milestone with a "**why now**" compliance snapshot or industry benchmark. This keeps vision tied to risk posture and external credibility, not just internal ambition.

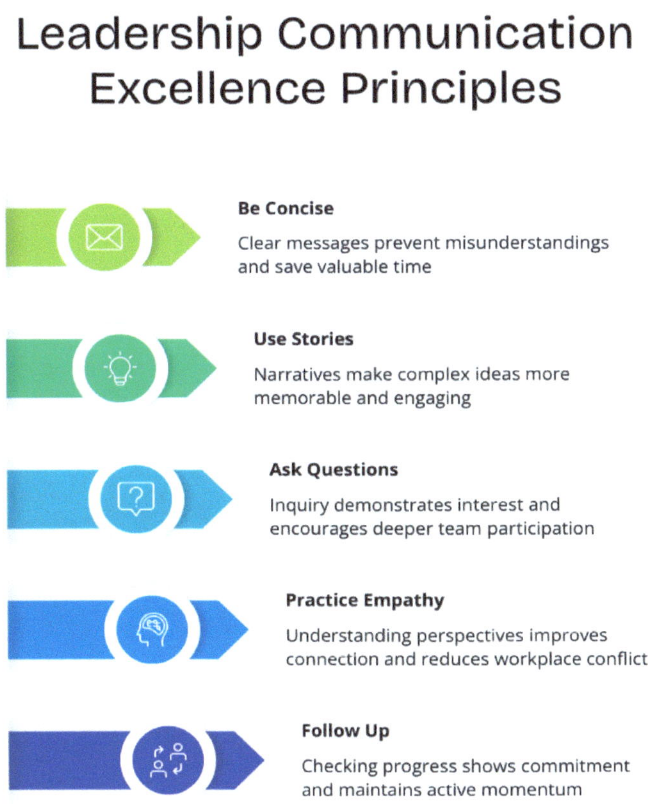

Execution: Turning Vision into Reality

The true measure of visionary leadership lies in execution. A well-crafted vision is only as valuable as its implementation. This requires:

- Meticulous planning,
- Milestone clarity,
- Resource allocation,
- Constant recalibration.

Break down the vision into **mission slices** 6-to-12-week outcome blocks with clear goals, owners, and success criteria. Avoid massive waterfall roadmaps. Instead, work in increments that demonstrate real user or business value.

Each mission slice should result in something visible: a new capability shipped, a pain point eliminated, or a decision unblocked. Execution is momentum made observable.

Build in fast feedback. Use metrics to track adoption, impact, and experience shifts. Establish a cadence:

- **Weekly**: blockers and clarity.
- **Monthly**: progress vs outcomes.
- **Quarterly**: vision recalibration.

Use retrospectives not just to fix delivery issues, but to stress test the vision. Ask: **is this still the right ambition? Is the customer still asking for this?**

Run **pre-mortems** before each major phase: "**If this failed six months from now, what would we have ignored?**" This builds contingency literacy and eliminates blind spots early.

Field-tested tool: in distributed teams, use a shared **Vision Operating Canvas**, a single sheet that maps outcomes, current state, risks, and success metrics. Refresh it monthly. Make it the heartbeat of every planning ritual.

Legacy and Long-Term Impact

The ultimate measure of a visionary IT leader lies not in immediate accomplishments but in what survives and evolves beyond them.

Legacy is built through:

- Durable **culture**, not transient performance.
- Sustainable **systems**, not short-lived wins.
- Empowered **people**, not dependent teams.

Create playbooks others can pick up. Document frameworks, toolkits, naming conventions, feedback loops, governance protocols, not for compliance, but so your vision can scale without you.

Test legacy readiness by stepping back for a sprint. **What breaks? What slows down? Where do decisions stall?** That's where your leadership hasn't been scaled yet.

Mentorship is not optional. Build second-line leaders who carry vision, not just execution. Use:

- **Shadow cycles,**
- **Vision brief co-authoring,**
- **Delivery team rotations.**

Your legacy will not be the tech stack you modernized. It will be the minds you elevated, the systems you embedded, and the beliefs you normalized.

Design for handoff, not headlines. Build what lasts. Teach what sticks. Empower what grows without you.

Strategic step: before closing out any major initiative, run a **handoff simulation**. Ask a rising leader to step in, direct the final 10% delivery, and own the post-launch ops. You'll learn instantly what's clear, what's missing, and where legacy still needs scaffolding.

The Journey Begins Here

Every leader's story starts not with a title, but with a decision, to act differently, to think longer, and to lead forward. Visionary IT leadership is not a destination. It is a pattern of behavior, a mindset of pattern recognition, and a refusal to settle for maintaining what already exists.

The journey ahead is not just about what technology can do, it's about what you choose to make it do. It's about how you align people, decisions, and systems toward a shared future that doesn't exist yet, until you build it.

This chapter was not written to inspire. It was written to equip.

You now have the tools to ask better questions, set clearer direction, shape lasting culture, and deliver transformation that outlives your position.

The future of IT will not be inherited.

It will be authored.

And that authorship starts here: not with a quote, but with a question, what is one strategic decision you'll take this week that reflects the vision you haven't yet shared? Start there. Then lead forward.

The Journey Begins Here

Bold decisions, lasting impact, your leadership story starts now

Chapter 2

Governance: The Foundation of Trust

Governance is the backbone of effective IT leadership, an essential force that establishes trust, accountability, and resilience in the face of rapid technological change. Far from being a **bureaucratic** exercise, **governance** enables organizations to align their processes, decisions, and resources with strategic goals, ensuring that every action contributes to long-term success. For IT leaders, governance is the bridge between innovation and stability. It provides the clarity and structure necessary to navigate complex challenges while fostering an environment of trust and collaboration.

Governance should be designed like a control tower, not a toll gate. It should give teams and stakeholders visibility into priority signals, risk exposure, budget compliance, architectural dependencies. while enabling decisions, not slowing them.

Governance is not just about maintaining order or compliance; it's about enabling excellence. Embedding governance into the organizational culture enables IT leaders to create a framework where teams can thrive, stakeholders feel secure, and customers trust the integrity of the organization. This chapter delves into the critical role of governance, its foundational principles, and how IT leaders can implement it effectively to drive meaningful, sustainable outcomes.

In practice, this means integrating governance triggers directly into delivery workflows: change requests tied to version control, security reviews built into sprint planning, and executive checkpoints triggered by data thresholds, not calendar dates.

Treat these as **living guardrails**. They guide motion, not block it. Build them into your delivery tools, not separate committees, so governance becomes something you do, not something you attend.

The Strategic Importance of Governance

Governance in IT leadership is a strategic imperative that safeguards organizations from inefficiency, misalignment, and risk. It ensures that decisions are made responsibly, that resources are allocated effectively, and that processes are designed to withstand scrutiny. Governance builds confidence among stakeholders by demonstrating that the organization operates with integrity, foresight, and accountability.

In sectors like healthcare and finance, governance also enables compliance with strict auditability, think HIPAA, GDPR, or SOX while still supporting innovation. Smart governance embeds these requirements in reusable templates, project charters, and API contracts.

Build compliance as code, automated validation scripts, standardized privacy checklists, and real-time dashboards that flag exceptions before audits do.

Trust is the cornerstone of governance. For teams, governance means that roles, responsibilities, and expectations are clearly defined, enabling collaboration and focus. For stakeholders, it provides assurance that their investments are being managed wisely and aligned with the organization's objectives. And for customers, governance ensures that their data, privacy, and experience are treated with the utmost care and responsibility.

To operationalize trust, leaders should define a shared "trust contract" what the organization commits to protect (data, uptime, experience), how it will signal adherence, and how issues will be escalated or corrected in transparent ways.

Imagine an IT initiative launched without governance. Unclear ownership, mismanaged risks, and non-compliance with regulations could derail even the most promising project. Governance prevents these pitfalls by providing a structured approach to decision-making and execution. It transforms uncertainty into opportunity and ensures that every initiative contributes to the broader mission of the organization.

Run **governance readiness** reviews before kickoff: confirm data-handling owners, budget authorization, and policy alignment. Treat readiness as the first deliverable.

GOVERNANCE PRINCIPLES FOR IT LEADERSHIP

The Foundations of Governance

Effective governance is rooted in principles that serve as its foundation. These principles ensure that governance is not only robust but also adaptable to the needs of the organization. Embracing these principles enables IT leaders to create a governance framework that drives efficiency, accountability, and trust. These principles should not sit in policy binders, they must be codified into day-to-day operations using decision logs, review cadences, and live dashboards that reflect them in action.

1. Transparency

Transparency is the foundation upon which governance is built. It ensures that all stakeholders' teams, executives, and customers have visibility into the decisions, processes, and outcomes of the organization. Transparency fosters trust by making the "**why**" and "**how**" behind decisions clear and accessible.

Do this: expose governance artifacts in real time, roadmaps, spend dashboards, audit findings, release calendars. When people can see the data, they stop guessing the intent.

Build transparency into tooling, not just communication. Use dashboards to expose service-level metrics, backlog health, and financial pacing. Use versioned documentation for architectural changes or compliance mapping. Transparency becomes real when it's persistent and retrievable.

To cultivate transparency, IT leaders must prioritize open communication and information sharing. This involves creating channels where key metrics, policies, and updates are readily available. For example, dashboards can provide real-time insights into project progress, budget utilization, and performance metrics.

Regular updates, whether through town halls or reports, ensure that everyone remains informed and aligned with organizational goals.

Adopt **show-the-work** Fridays, brief 15-minute team demos of what changed, what decisions were made, and what data supported them.

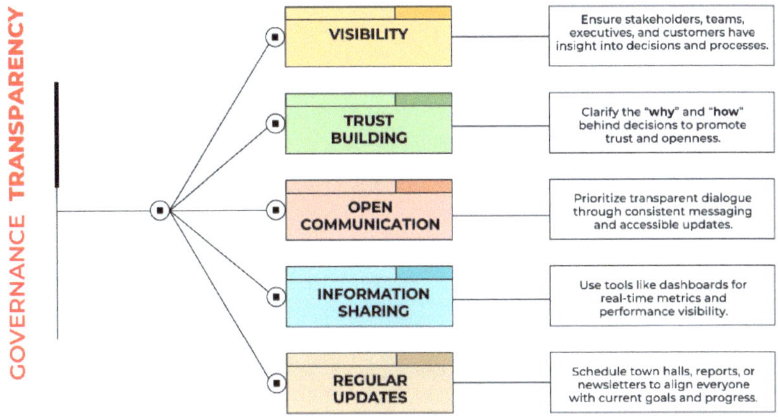

2. Accountability

Accountability ensures that every decision and task has a clear owner. It empowers individuals to take responsibility for their roles, fostering a sense of purpose and commitment. In governance, accountability drives performance by making sure that objectives are met with precision and integrity.

A robust way to institutionalize accountability is through automated logs and ownership tags within tools like service management platforms, change request systems, or budget allocations. Every action should be traced back to a decision-maker.

Establishing accountability requires clarity in roles and responsibilities. Frameworks like RACI provide structure, but they only work if roles are visible and reviewed.

Publish a **decision ledger** that lists each major initiative, its accountable owner, and last decision date. This becomes the organization's source of truth for accountability.

Quarterly accountability reviews should focus on what was delivered, what dependencies were managed, and what lessons were documented, not just metrics.

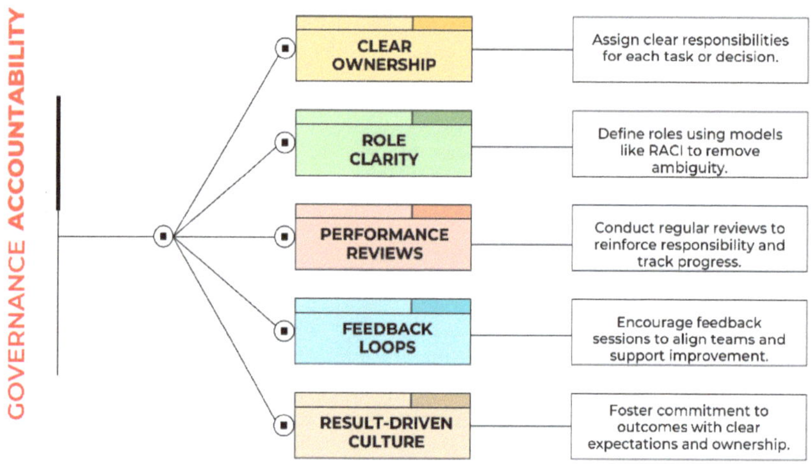

3. Risk Management

Risk is an inherent part of IT operations, but governance transforms it into a managed discipline. Identifying, assessing, and mitigating risks proactively ensures that organizations handle challenges without disruption.

Build an automated **risk scoring engine** tied to new initiatives. Every procurement, integration, or data use triggers an instant risk index, technical, legal, and financial.

Maintain a living risk register and refresh it quarterly. Link every risk to a mitigation plan, a responsible owner, and a review trigger.

Public-sector or utility environments use threat matrices that link risk to service obligations, adopt the same thinking for business continuity across all industries.

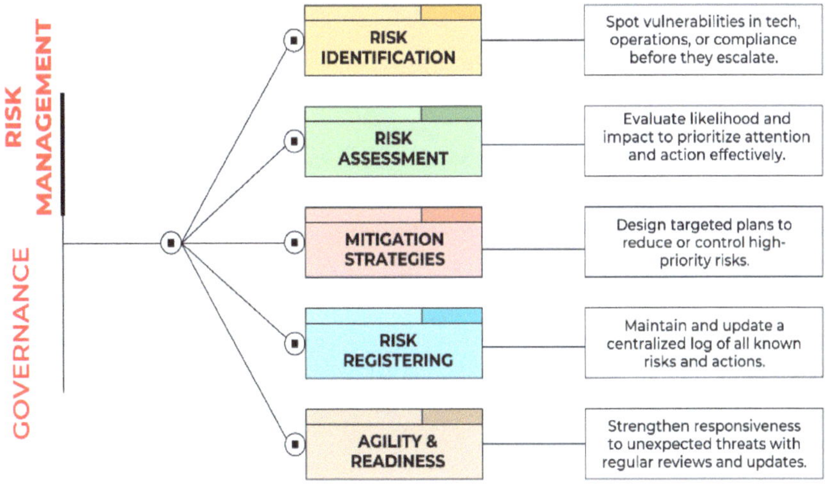

4. Ethical Decision-Making

Ethical considerations are central to governance. In a digital age defined by data and algorithms, governance must safeguard fairness, transparency, and accountability.

Operationalize ethics like security: include an **Ethical Impact** field in every change request, who could be harmed, excluded, or surveilled by this decision?

Establish ethics review check-ins for initiatives involving AI, automation, or data profiling. These are not obstacles; they're confidence builders for executives and customers alike.

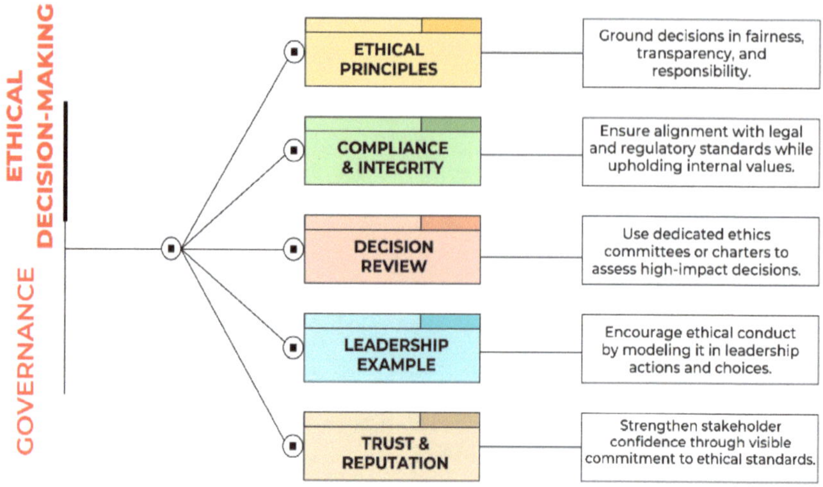

Designing and Implementing Effective Governance

Implementation is where governance becomes tangible. It must be treated like an embedded service layer, visible in every workflow, reusable across teams, and continuously improving.

Define scope and objectives first. Decide which domains, compliance, risk, or performance, need governance most urgently. Then choose and customize frameworks (COBIT, ITIL, NIST) to fit your organization's reality.

Do this: Map each governance domain to a measurable output e.g., reduced audit findings, faster change approvals, or fewer SLA breaches. Governance earns credibility through results, not reports.

Integration means inserting governance into everyday systems. Embed stage-gates into DevOps pipelines, automate audit checklists, and connect performance metrics directly to leadership dashboards.

Communication is vital. Leaders must demystify governance through open workshops, office hours, and quick-start guides.

Launch a **governance playroom,** a shared space (physical or digital) where teams can test processes, suggest simplifications, and see how rules evolve.

Maintain a live governance knowledge hub: quick templates, FAQs, decision trees, and progress trackers. This converts governance from theory into enablement.

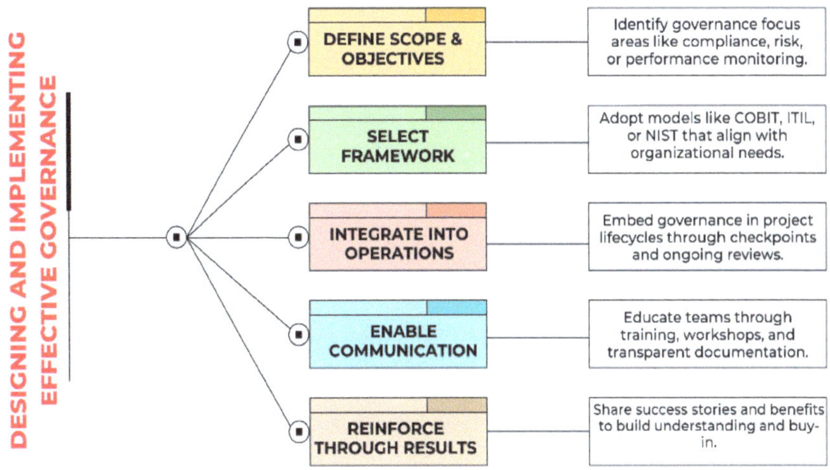

Overcoming Challenges in Governance

Resistance, complexity, and misconceptions are common hurdles, but they can be reframed as design inputs.

Resistance often stems from fear that governance restricts autonomy. Counter this by proving that good governance accelerates delivery.

Run side-by-side pilots: one project with governance embedded, one without. Measure delivery time, rework, and risk outcomes. Share the data, it will convert skeptics faster than memos.

Start small. Focus governance in one high-impact area before scaling. Simplify. Iterate. Build proof of value.

Misconceptions thrive in silence. Keep governance visible. Communicate why each rule exists, who it protects, and how it reduces waste.

Gamify adoption. Reward teams that close audits with zero findings or that reduce approval cycle time via governance automation.

The Long-Term Value of Governance

When governance is implemented effectively, it transcends compliance, it becomes culture. It shapes how teams collaborate, how stakeholders make decisions, and how customers perceive trust.

Imagine teams operating with clear boundaries, decisions visible, and shared accountability. Stakeholders are confident in every investment. Customers are loyal because transparency breeds reliability. That is what enduring governance looks like.

Governance maturity can be measured. Track three signals quarterly: number of exceptions caught early, number of escalations resolved within SLA, and percentage of automated compliance checks.

A well-architected governance model survives leadership turnover, mergers, and crises because it is principle-based, not personality-driven.

Building a Legacy Through Governance

Governance is more than a framework, it is a leader's legacy. When embedded properly, it becomes a system of trust that endures beyond your tenure.

To sustain it, conduct annual **governance retrospectives** with independent reviewers. Ask: Is it still enabling speed? Is it still protecting value? Is it still ethical?

Effective governance evolves. It must flex with new technologies, regulatory climates, and organizational growth.

In governance lies the promise of clarity, accountability, and ethical leadership. It is the framework through which IT leaders inspire trust, drive excellence, and shape the future.

Governance is your invisible superpower, quiet when it works, essential when pressure rises, and remembered when legacy is measured.

Keep it alive. Audit it. Simplify it. Celebrate it. Governance that breathes is governance that lasts.

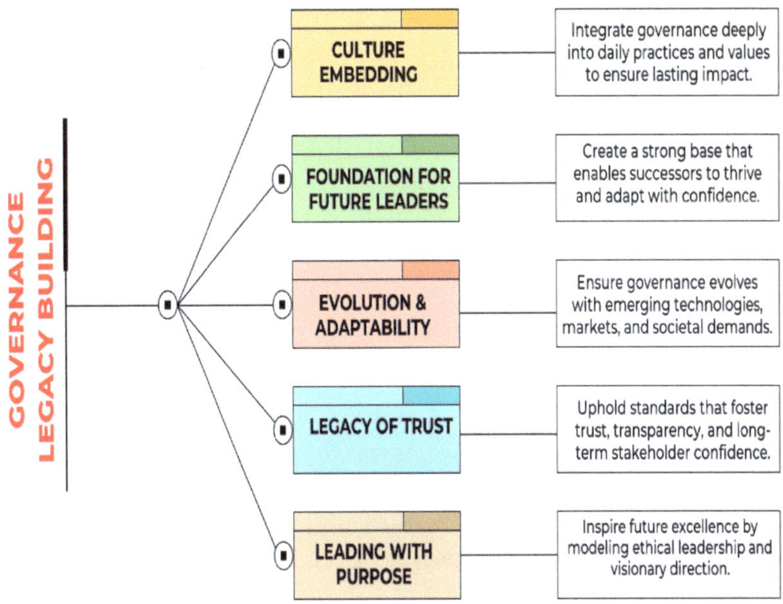

Chapter 3

Mastering Digital Transformation

Digital transformation defines the trajectory of progress in today's interconnected and technology driven world. It has become the foundation upon which organizations redefine their operations, compete in global markets, and create meaningful connections with their customers. While it often appears synonymous with adopting advanced technologies, it is much deeper, it requires rethinking processes, reimagining objectives, and reshaping organizational mindsets. This is a holistic journey of innovation, adaptation, and leadership.

In practical terms, digital transformation means upgrading decision-making speed, automating where it counts, and realigning delivery teams with customer outcomes, not just digitizing paper workflows.

Digital transformation is not a program, it is a managed shift in how an organization operates. It must be run with discipline, but measured by experience, not artifacts.

To master digital transformation is to see it as a tapestry where technology, people, and strategy intertwine. Success comes when these elements work in harmony, each reinforcing the other. In this chapter, we will explore **how leaders can embrace this journey, turn challenges into opportunities**, and position their organizations to thrive in an environment defined by constant change and technological evolution. Use transformation management offices (TMOs) not as reporting centers, but as orchestration hubs that translate strategy into multi-track delivery.

Organizations must treat transformation as an operating model upgrade, not a technology refresh, structured with defined outcomes, realignment cadences, and change management rhythms.

Think of quarterly release cycles, with each wave delivering measurable change in process efficiency, experience quality, or time-to-decision.

Establishing the Foundation for Transformation

The starting point for any digital transformation effort is a clear understanding of what it seeks to achieve. For organizations, this means moving beyond surface level improvements to address fundamental questions about their goals, values, and aspirations. The most successful transformations are built on a well-defined purpose that serves as the guiding principle throughout the journey. A lack of clarity at this stage can lead to fragmented efforts, inefficiencies, and missed opportunities.

The foundation should be captured as a 'transformation charter' a one-page artifact that outlines the business outcome, metrics of success, affected domains (processes, roles, tech), and stakeholder sponsors. This becomes the reference point when priorities shift or scope creeps.

Treat this charter like a living contract: revise it quarterly as learnings emerge, and publish updates visibly to all delivery teams.

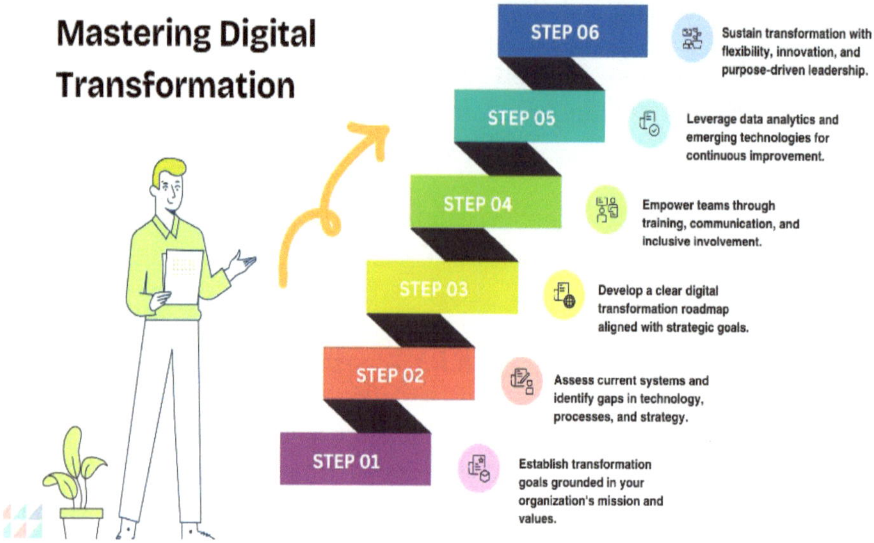

The foundation for transformation begins with an honest assessment of the organization's current state. Leaders need to analyze their processes, technologies, and customer relationships to identify gaps and opportunities. This step is critical because it allows organizations to align their digital initiatives with their overarching mission. Whether the aim is to improve customer experience, streamline operations, or unlock new revenue streams, every decision must be anchored in a vision that resonates across the organization.

Use diagnostics like capability maturity assessments, value stream maps, or system lifecycle health scans. In financial services, these tools help quantify friction in lending processes or back-office automation. In healthcare, they expose latency between patient engagement and outcomes.

In manufacturing, latency mapping of shop-floor to ERP updates can identify digital thread gaps that slow decisions.

Clarity of purpose also enables alignment between teams and stakeholders. When everyone understands why transformation is necessary and what it aims to achieve, collaboration becomes more focused and impactful. This shared sense of purpose fosters trust and engagement, making it easier to navigate the complexities that inevitably arise during the transformation process.

To codify alignment, run vision-to-value workshops with cross-functional leaders, mapping each initiative to a business driver and assigning shared KPIs. This prevents siloed projects that miss enterprise priorities.

Use joint OKRs across business and IT leaders so transformation outcomes are co-owned, not just delegated.

Leadership and Vision in the Transformation Journey

Leadership plays a pivotal role in digital transformation. It is the leader's responsibility to articulate a vision that inspires, motivates, and unites teams. A compelling vision is not just aspirational, it is actionable. It outlines what the future could look like and provides a clear path to get there. Leaders must communicate this vision with conviction and authenticity, ensuring that it resonates with employees, stakeholders, and customers alike.

An actionable vision also includes "**what we'll stop doing**" and "**how we'll measure change.**" This creates traction, not just inspiration. In manufacturing, this might mean sunsetting manual inspection steps and linking vision to product cycle acceleration.

In retail, this might mean eliminating legacy POS workflows in favor of mobile-first checkout linked to supply chain telemetry.

The process of crafting this vision begins with understanding the broader context in which the organization operates. Leaders need to evaluate the competitive landscape, identify emerging trends, and anticipate how market dynamics are likely to evolve. This foresight allows them to position their organization strategically, ensuring that transformation efforts address not only immediate challenges but also future opportunities.

Use external benchmarking, customer voice analysis, and trend signal tracking. In the public sector, this could involve cross-agency insights about digital ID usage, in retail, changes in customer mobile behavior. Use these to build a 'next normal' playbook, not just a tech roadmap.

In utilities, this may mean tracking smart grid adoption rates and policy shifts in carbon credits to anticipate digitization priorities.

Equally important is the ability to translate vision into strategy. A well-defined strategy bridges the gap between ideas and execution. It prioritizes initiatives, allocates resources effectively, and establishes a roadmap for achieving transformational goals. Leaders must also remain flexible, adjusting their strategy as new information emerges or circumstances change. This adaptability is a hallmark of successful transformation leadership.

Structure strategy execution as a rolling portfolio 12-week cadence reviews, with stage gates tied to value metrics and dependency health. In insurance and utilities, this allows adjustment when policy shifts or new regulations emerge.

Use **initiative health boards** that color-code programs not just by budget, but by stakeholder confidence and friction reduction achieved.

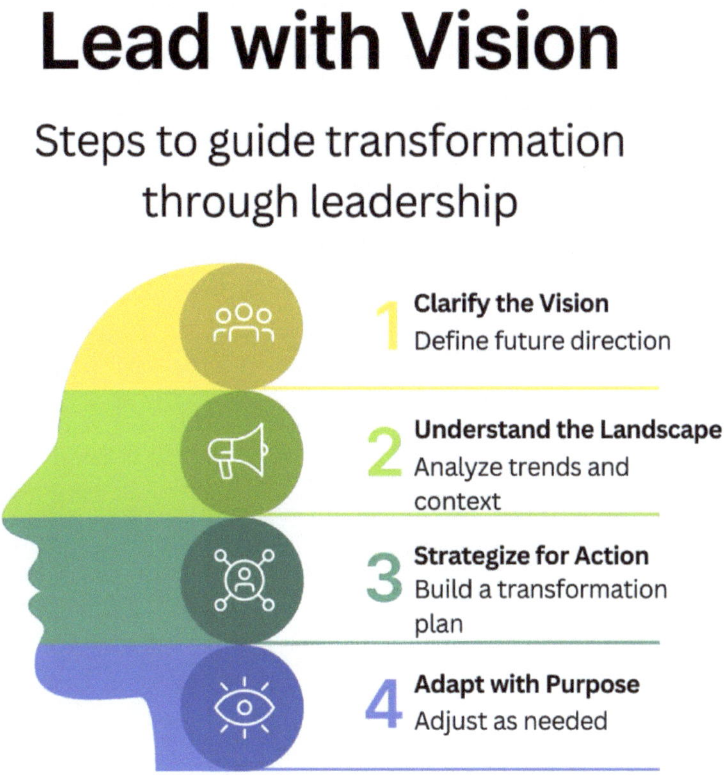

Empowering People to Drive Change

At the heart of every successful digital transformation are the people who make it happen. Employees, stakeholders, and customers play a vital role in shaping the outcomes of transformation efforts. Recognizing the importance of the human element is essential for leaders who want to create meaningful and lasting change. After all, the most sophisticated technologies and strategies are only as effective as the people who implement and support them.

Transformation is a change in behavior, not just systems. Organizations must treat people enablement as an active workstream, with defined goals, behavioral shifts to track, and feedback cycles tied to business results.

Launch behavior adoption metrics: measure uptake of self-service tools, digital-first process usage, or handoff accuracy.

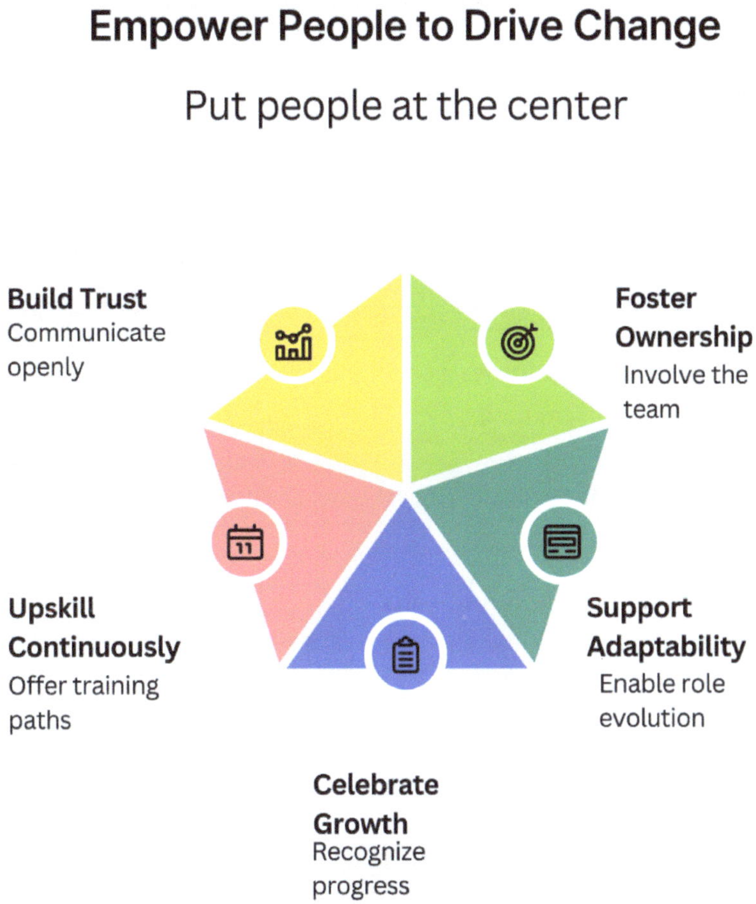

One of the most significant challenges in transformation is overcoming resistance to change. People naturally gravitate toward familiar processes and routines, which can make them hesitant to embrace new ways of working. To address this, leaders must prioritize open communication and transparency. Explaining the rationale behind transformation initiatives and involving employees in the decision-making process, leaders can build trust and foster a sense of ownership.

Run **change readiness** assessments by department, mapping emotional drivers and barriers. Use that data to design engagement campaigns, not generic rollouts. In logistics or manufacturing, this may include shift-based training, job shadowing for digital tools, or live demos tailored to floor-level impact.

Use **change infographics** to summarize before/after for every process, make the benefit visual, not abstract.

Investing in the development of employees is another critical aspect of empowering people. Digital transformation often requires new skills and competencies, particularly as organizations adopt advanced technologies and data driven approaches. Leaders should provide opportunities for training, mentorship, and continuous learning, enabling employees to adapt to their evolving roles. This not only boosts morale but also enhances the organization's overall capacity for innovation and resilience.

Structure this as a workforce enablement plan with three layers: foundational (digital fluency), functional (role-specific tools like ERP or CRM), and strategic (cross-domain thinking like design thinking or data literacy). Assign transformation mentors to support frontline teams.

Track learning adoption by role and correlate with process improvement metrics, this builds the case for continued investment.

Leveraging Technology for Strategic Impact

Technology is the catalyst for digital transformation, but its value lies in how it is used to achieve strategic objectives. Leaders must approach technology with a clear understanding of its potential to drive innovation, efficiency, and growth. This requires thoughtful planning, deliberate implementation, and a commitment to continuous improvement.

Every technological investment must answer: **how does this reduce friction, increase speed to value, or improve experience for users or customers?** Build a tech prioritization matrix around these criteria to filter initiatives.

Use a **value-to-effort** quadrant: prioritize high-impact, low-complexity tech moves (e.g., workflow digitization) before heavy investments like platform overhauls.

The first step in leveraging technology effectively is assessing the organization's current infrastructure. Leaders need to identify outdated systems, inefficiencies, and areas for improvement. This evaluation should be guided by the organization's strategic goals, ensuring that technology investments align with broader priorities. For example, adopting cloud-based solutions may enhance flexibility and scalability, while integrating artificial intelligence can improve decision making and customer experience.

Conduct a **tech debt heatmap** by capability, overlayed with business value impact. In government, this might surface mainframe processes that delay licensing, in education, manual scheduling tools. Tie the assessment to current customer friction points.

Map **tech-to-outcome chains** showing how each system directly or indirectly supports value streams, so decisions are based on business fit, not tech nostalgia.

Innovation is central to harnessing technology's full potential. Leaders should encourage experimentation and exploration, creating an environment where new ideas can flourish. Pilot programs allow organizations to test emerging technologies on a small scale before scaling successful initiatives. This iterative approach minimizes risks while maximizing impact, enabling organizations to stay ahead of technological trends.

Use a formalized tech incubation framework. Define sandbox environments, set a 90-day pilot cycle with exit criteria, assign business sponsors, and include readiness assessments before scaling. Public sector teams can test AI-assisted document triage; healthcare can trial chatbots for triage routing.

Tie each pilot to a measurable hypothesis **reduce intake time by 30%** and build a post-pilot decision rubric: scale, shelve, or shift.

Data also plays a critical role in digital transformation. Leveraging analytics and insights, organizations can make informed decisions, optimize processes, and deliver personalized experiences. For instance, predictive analytics can help businesses anticipate customer needs, while machine learning algorithms can uncover patterns and opportunities that might otherwise go unnoticed.

Build a data-to-decision pipeline: define which decisions need real-time vs. periodic data; set up data owners and stewards per domain; and embed data visualization into planning, not just reporting. In retail, this enables personalized inventory allocation. In healthcare, it supports risk stratification for patient care.

Operationalize data literacy: ensure all teams can interpret dashboards tied to their KPIs, and build data question prompts into team retrospectives.

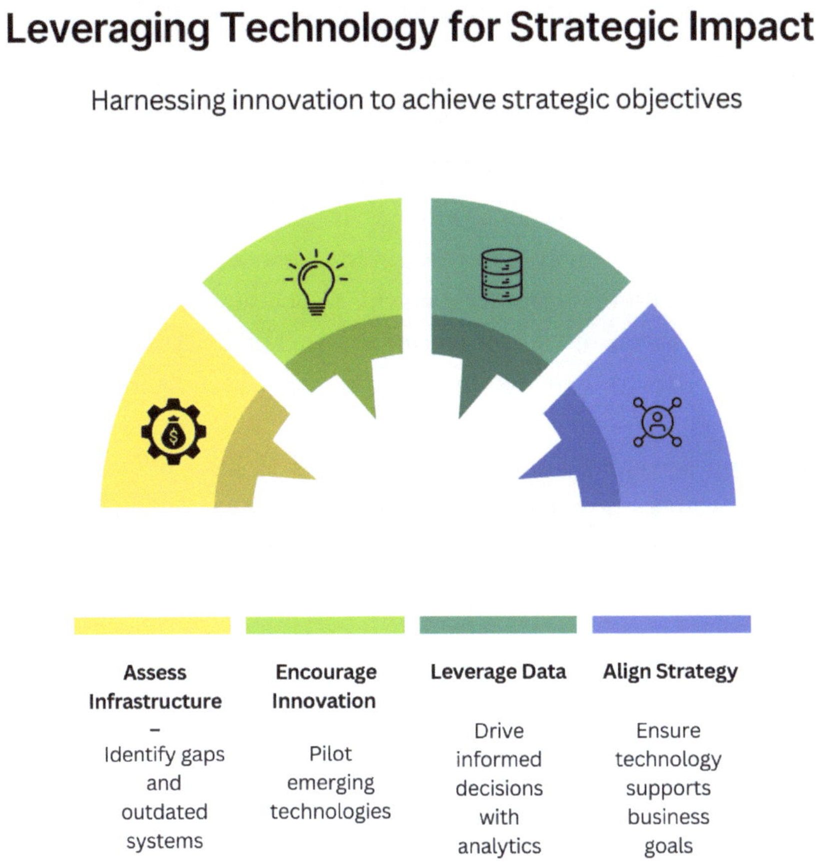

Redefining Customer Experience

In the digital age, customers expect seamless, personalized, and intuitive interactions. Meeting these expectations requires a shift in focus from products and services to the overall customer experience. Digital transformation offers organizations the tools and strategies to deliver on these expectations, creating deeper connections and building lasting loyalty. To redesign experience, create cross-functional customer journey maps that span onboarding, service delivery, and retention. Layer operational data and customer sentiment onto these maps to isolate pain points by channel, time, or persona.

Use CX triangulation: match user verbatims, system logs, and service desk tickets to find root experience blockers.

Understanding customer journeys is a foundational step in transforming customer experience. Organizations need to map out every interaction a customer has with their brand, identifying pain points and opportunities for improvement. This holistic view allows leaders to design experiences that are not only efficient but also emotionally resonant. In financial services, this may involve digitizing account opening with embedded identity verification. In public agencies, it could mean reducing paperwork touchpoints in permit issuance. Use **moment-of-truth** analysis to focus redesign critical impact steps.

Score each interaction on effort, emotion, and expectation gap, prioritize redesigns that hit all three.

Technology enables organizations to tailor experiences to individual customers. For example, advanced analytics can provide insights into customer preferences and behaviors, allowing businesses to deliver personalized recommendations, offers, and support. Omnichannel strategies further enhance customer experience by ensuring consistency across multiple touchpoints, whether online, in store, or through mobile platforms.

Use customer data platforms (CDPs) to centralize interaction data. From there, set up dynamic segmentation rules to trigger real-time personalization, automated emails, app notifications, or next-best-action suggestions. For manufacturers, this could guide post-sale support outreach based on IoT data.

Apply **predict-next-need** modeling to proactively serve users based on historical sequences, reducing drop-offs and improving satisfaction.

Prioritizing customer centricity, organizations can create experiences that delight and differentiate. This focus not only drives customer satisfaction but also strengthens the organization's competitive position in an increasingly crowded marketplace.

Establish customer experience metrics like Net Promoter Score (NPS), First Contact Resolution (FCR), or Customer Effort Score (CES), and align them to business outcomes such as churn reduction or upsell conversion. Build these into team OKRs to make customer value visible across functions.

Use voice-of-customer heatmaps to track experience friction by geography, segment, or channel, and route data to relevant teams weekly.

Navigating Challenges and Sustaining Momentum

Digital transformation is a complex and multifaceted journey, and challenges are inevitable. From budget constraints to legacy systems, organizations must navigate a range of obstacles to achieve their goals. However, with careful planning and strategic problem solving, these challenges can be transformed into opportunities for growth.

Treat transformation as a managed portfolio with risk-adjusted funding, multi-speed delivery models, and embedded escalation routes. Create a central transformation PMO to steer governance, track dependencies, and align effort with outcomes.

Use quarterly transformation reviews with **red/yellow/green** indicators by domain, focusing executive attention on friction, not status.

One common challenge is the inertia of legacy systems. Outdated technologies can hinder progress, create inefficiencies and limit the organization's ability to innovate. Addressing this requires a phased approach that balances modernization with continuity. Leaders should prioritize critical systems for immediate upgrades while gradually transitioning fewer essential components.

Use application rationalization tools to score each system on business criticality, integration complexity, and total cost of ownership. Build phased migration playbooks: lift-and-shift for infrastructure-heavy apps; rebuild for customer-facing digital tools; retire or replatform where risk is low.

Tie modernization to business cycle moments e.g., fiscal calendar, seasonal loads, or regulatory timelines, to avoid disruption.

Financial constraints are another hurdle, particularly for organizations operating in resource constrained environments. To overcome this, leaders must focus on high impact initiatives that deliver measurable value. Partnering with external vendors or leveraging open-source solutions can also help organizations achieve their goals without overextending their budgets.

Apply a **value stream funding** model, allocate budget by business outcome, not department. Use gated release of funds based on proof-of-value or velocity milestones. This method is effective in healthcare networks, where payer-model constraints demand constant reprioritization.

Link funding gates to transformation impact scores, not just activity logs, this ensures ROI visibility at every stage.

Resistance to change remains a significant challenge, particularly when it comes to shifting organizational culture. Leaders must address this resistance proactively by fostering a culture of openness, collaboration, and experimentation. Celebrating milestones and successes can also build momentum, encouraging employees to embrace the transformation journey.

Use quarterly pulse checks to track sentiment, stress indicators, and engagement levels. Set up transformation champions across departments to serve as culture catalysts and local coaches. Make small wins visible and measurable across the enterprise.

Turn champions into **feedback amplifiers** collect real-time blockers and push insights into the transformation operating rhythm.

The Lasting Impact of Transformation

Digital transformation is not a one-time initiative it is a continuous process of adaptation and evolution. Organizations that embrace this mindset are better equipped to navigate the uncertainties of a rapidly changing world, positioning themselves as leaders in their industries. The benefits of transformation extend far beyond operational improvements; they create a foundation for innovation, resilience, and growth.

Build a feedback loop into your operating rhythm, review transformation impact quarterly using leading indicators (speed, customer satisfaction, digital adoption) and lagging indicators (revenue lift, cost-to-serve, retention). Use these reviews to refresh strategy dynamically.

Create **impact dashboards** visible to all levels, not just execs, so transformation feels shared, not distant.

Leaders who master digital transformation leave a lasting legacy. Their efforts not only shape the organization's present but also influence its future trajectory. Aligning technology with purpose, engaging people with authenticity, and overcoming challenges with determination, leaders can create an environment where progress is not only possible but inevitable.

Sustained transformation depends on succession planning. Document what's repeatable, train teams in transformation literacy, and hardwire agility into governance, not just strategy.

Make your playbooks modular, reusable, and tool-agnostic, so any team, anytime, can pick them up and execute

The journey of digital transformation is as much about mindset as it is about strategy. It requires vision, courage, and a commitment to continuous improvement. For organizations willing to take this journey, the rewards are profound greater agility, deeper customer relationships, and a legacy of innovation that sets them apart.

Transformations that endure are those rooted in operational excellence and driven by purpose, not trend chasing. Make purpose measurable, and excellence repeatable.

Document **why this worked** at the end of each phase, not just what was built, so others can repeat without reinventing.

Mastering digital transformation is a calling for those who dare to lead boldly and think creatively. As you embark on this journey, remember that every step forward brings you closer to your goals. The future is being shaped by those who embrace transformation with purpose and passion. This is your opportunity to lead the charge, inspire change, and leave an indelible mark on your organization and the world.

This is not just your opportunity, it is your obligation to architect change that scales, endures, and earns trust.

Chapter 4

Building a Secure Digital Fortress

In an era where cyber threats evolve at an unprecedented pace, securing your organization's digital assets is more than a responsibility, it is a strategic imperative. A secure digital fortress is not limited to implementing advanced technology; it is about fostering a culture of vigilance, adapting to emerging challenges, and embedding trust into the very fabric of your organization.

Security must be embedded into business planning, procurement, software design, data governance, and user experience, not siloed as a reactive IT function. Build your approach using the **secure by design** model, treating risk mitigation as an upfront investment rather than a downstream fix.

The stakes are high. A single breach can erode years of customer trust, disrupt critical operations, and result in severe financial and reputational damage. Security is more than a reactive measure; it is an active force that empowers your organization to innovate fearlessly while safeguarding its most valuable assets.

Embed secure design reviews into product development. Before launching any service, run a security threat model that evaluates access points, data flow, and abuse scenarios. Use security story mapping alongside user journeys to visualize protection needs.

Building a secure digital fortress, you are not just protecting information; you are demonstrating your commitment to those who depend on you, employees, customers, and partners alike.

In healthcare or finance, where regulatory and privacy stakes are amplified, align architecture with frameworks like NIST CSF or ISO/IEC 27001. Perform a crosswalk mapping of controls to business workflows and data lifecycles.

This chapter lays out the strategies, tools, and cultural shifts required to construct and maintain a fortress that not only resists threats but adapts and evolves to face the challenges of tomorrow.

Why Digital Security is the Bedrock of Success

Security is integral to every facet of IT leadership. It enables organizations to operate without disruption, build trust among stakeholders, and pursue growth with confidence.

Conduct security reviews in parallel with strategic planning. For every major program or product rollout, assign a **risk champion** to identify vulnerabilities early and report mitigation actions back to executive steering

Without strong security measures, even the most advanced digital transformation efforts are vulnerable to collapse.

For each program initiative, create a security risk baseline. Score likely threat vectors using CIA (Confidentiality, Integrity, Availability) impact and assign a mitigation owner. For public-facing services, mandate pre-launch penetration testing.

A proactive approach to security fosters confidence, not only within your organization but also across your network of stakeholders.

Design business continuity blueprints based on operational criticality. For example, assign the highest protection levels to systems like patient EHRs (healthcare), clearinghouses (finance), or voting systems (government).

When customers know their data is handled securely, they are more likely to engage with and trust your brand. When employees feel equipped with the tools and training to uphold security, they operate with greater confidence and efficiency.

The Core Pillars of a Secure Digital Fortress

Every robust security strategy rest on several essential components: **governance, technology, people, and adaptability**. Together, they form an ecosystem designed to withstand evolving threats while enabling the organization to operate efficiently and effectively.

Governance: Rules and Responsibility

Governance establishes the framework for how security is implemented, monitored, and improved.

Create a Security Governance Charter. Include ownership of threat detection, response coordination, policy lifecycle management, and regulatory tracking. Ensure visibility by logging into all decisions in a version-controlled playbook.

Setting clear policies and assigning accountability, governance ensures that security becomes an integral part of daily operations rather than an afterthought.

Begin by defining comprehensive security policies tailored to your organization's needs. These policies should address areas such as data access, password management, and incident response.

Use a central policy repository, connected to your workflow platforms, where policy changes trigger alerts to relevant owners. Set up dynamic policy review cycles (e.g., every 180 days) instead of relying on annual reviews.

Align these with global regulations like GDPR, HIPAA, or CCPA to ensure compliance. Documenting these policies and making them accessible fosters consistency and clarity.

Accountability is the cornerstone of effective governance.

Use governance matrices to assign roles to specific controls and triggers. For example, finance systems should have designated approvers for payment integrity alerts; cloud systems should log access elevation requests with timestamps and approver ID.

Establish ownership for specific security initiatives and ensure escalation protocols are in place for handling potential breaches or risks.

Technology: The Tools of Defense

Technology serves as the foundation for protecting digital environments, but its effectiveness lies in how it is deployed and integrated.

Start with a Security Architecture Blueprint. Map controls across these planes: (1) Identity & Access, (2) Network & Perimeter, (3)

Application Layer, (4) Data Lifecycle. Use this blueprint to identify control gaps.

A well-constructed security architecture uses multiple layers of defense to mitigate risks at every level.

In manufacturing, deploy intrusion detection at the OT layer to prevent SCADA-based attacks. In finance, instruments all high-value transactions with anomaly detection and logging triggers.

To address more sophisticated threats, integrate advanced solutions like AI-powered threat detection.

Encryption plays a crucial role in ensuring sensitive data remains secure.

Use granular encryption key policies. For regulated sectors, rotate keys every 90 days and log access attempts. Align encryption tiers to data classification, confidential vs. restricted vs. public.

Ensure resilience through backup and disaster recovery systems.

Document RTO and RPO targets by business line. In cloud-native setups, automate backup validation tests (e.g., simulate full restores biweekly). Test failover scenarios across primary and secondary regions quarterly.

People: The Human Element of Security

Employees are often the first line of defense, and the most vulnerable link in the security chain.

Develop a security awareness matrix per role. Each function (engineering, sales, customer support) should receive training customized to their risk exposure. Assign mandatory simulations quarterly, not annually.

Security awareness should not be treated as a one-time initiative.

Track phishing simulation KPIs: click rate, report rate, time-to-report. Use these metrics to assign refresher modules dynamically. Deploy microlearning (3-minute modules) accessible on mobile.

Leaders must lead by example.

Implement security behavior tracking in leadership dashboards. Display MFA adoption, last password rotation, and unencrypted storage detection by department.

Adaptive Security: Responding to Evolving Threats

The digital landscape is in constant flux, with new vulnerabilities emerging every day. A truly secure digital fortress is adaptive.

Build a threat model repository: catalog previous incidents, emerging threat vectors, and simulations. Schedule monthly reviews with security stakeholders to adjust controls and playbooks.

Zero Trust has emerged as a cornerstone of modern security strategies.

Roll out Zero Trust in phases: 1) Device posture verification, 2) Role-based access gating, 3) Continuous session validation. For remote-first teams, enforce app containerization and VPN split tunneling.

AI and automation are revolutionizing cybersecurity.

Use SOAR platforms to define auto-remediation workflows. Example: quarantine device → alert user → reset credentials → flag for human follow-up. Set severity thresholds based on data risk, not alert volume.

Continuous monitoring and real-time threat intelligence strengthen defenses.

Tie your SIEM alerts to business context. Flag anomalies based on normal usage baselines, not static rules. Example: A report pulled at 3AM from a different geo should trigger MFA + audit trail snapshot.

Building a Culture of Security

Security is not solely the responsibility of IT teams.

Make **security moments** a part of every all-hands: share a recent threat, walk through how it was mitigated, and recognize the team or individual responsible. Normalize talking about risk.

Fostering a security-first culture begins with transparency.

Recognize and reward contributions to security.

Launch a **Security Hero of the Quarter** recognition program. Publish contributions, like early vulnerability reporting or process improvements, on internal comms or newsletters.

Measuring and Sustaining Progress

A secure digital fortress is not static; it evolves.

Create a living Security KPI Dashboard. Include: number of incidents (closed/open), average response time, employee phishing simulation success, percent systems with MFA enabled, and audit readiness index.

Establish clear metrics to measure success.

Use real-time compliance scoring across domains, map PCI, HIPAA, SOC2 controls to dashboards and highlight deltas. Schedule rolling internal audits, not annual events.

Stay informed about developments in cybersecurity.

Host quarterly cross-industry threat briefings. Include a rotating panel from public sector (CISA), healthcare (H-ISAC), and finance (FS-ISAC). Distribute a "threat digest" with actionable insights every month.

The Path to Confidence and Trust

A secure digital fortress is more than just a protective barrier, it is a foundation of confidence.

Use public-facing **Trust Centers** to communicate your security posture. Include uptime metrics, audit certifications, and data handling FAQs. Provide links to report vulnerabilities via bug bounty programs.

When security is prioritized, trust naturally follows.

During major product launches or outages, proactively communicate how user data remains protected. Embed "secure-by-default" into customer messaging.

Organizations that treat security as a business advantage, rather than a compliance checkbox, are the ones trusted to lead through disruption.

Security is not the absence of threats; it is the presence of readiness. Let your systems, your people, and your decisions prove it.

You can learn more about cybersecurity strategies and solutions at the **American Accel** website: **American-Accel.com** American Accel is recognized as a trusted leader in delivering **24/7/365 protection** across the full spectrum of cybersecurity. The company's solutions are designed to help organizations strengthen defenses, reduce risk, and build long-term resilience. Core offerings include:

- SIEM (Security Information and Event Management): Real-time monitoring and rapid detection of suspicious activity.
- EDR (Endpoint Detection and Response): Continuous protection against advanced endpoint attacks.
- MDR (Managed Detection and Response): Around-the-clock monitoring and response delivered by security experts.
- XDR (Extended Detection and Response): Unified visibility and detection across endpoints, networks, and cloud environments.
- SOAR (Security Orchestration, Automation, and Response): Automated playbooks that accelerate incident response.
- SOC (Security Operations Center): Centralized, real-time threat monitoring and defense.
- Dark Web Monitoring: Constant scanning for exposed credentials and breach intelligence.
- Risk Assessments & Vulnerability Management: Identifying and addressing weaknesses before they are exploited.
- BCDR (Backup and Disaster Recovery): Ensuring resilience and rapid recovery from disruptions.
- Email Security: AI-powered protection against phishing, ransomware, and Business Email Compromise (BEC).
- RMM (Remote Monitoring & Management): Proactive system monitoring and automation to maintain security readiness.

In addition to these solutions, American Accel structures its services around the **NIST Cybersecurity Framework** Govern, Identify, Protect, Detect, Respond, and Recover providing organizations with a structured, standards-based approach to security.

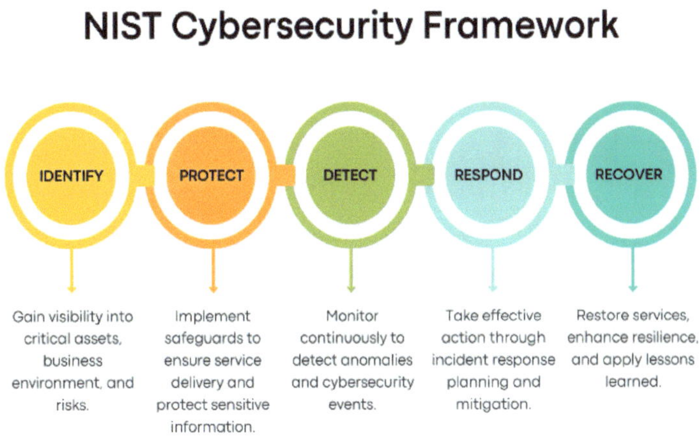

For a complete overview of services and to explore how American Accel supports businesses across industries, visit

American-Accel.com.

AMERICAN ACCEL

AI . CYBERSECURITY . TECHNOLOGY

Part 2

Leading Through Challenges (Chapters 5–7)

Part 2 delves into the critical role IT leaders play in navigating challenges and fostering growth, providing actionable strategies to thrive in demanding situations. **Chapter 5** focuses on leadership during IT crises, offering practical guidance to remain composed and decisive while managing emergencies effectively. It emphasizes building trust, deploying rapid solutions, and learning from crises to strengthen organizational resilience.

In **Chapter 6**, attention turns to building legendary teams, highlighting the importance of cultivating talent, fostering collaboration, and nurturing high performing groups. Leaders are encouraged to create inclusive environments that empower individuals to contribute their best and drive collective success.

Chapter 7 examines IT's role as a catalyst for business growth, showing how to align technology strategies with organizational goals to unlock innovation and expand opportunities. Leaders are guided to position IT as a strategic partner that adds measurable value across the enterprise.

This section equips IT leaders to lead with confidence, transform challenges into opportunities, and create pathways for sustained growth and innovation.

Chapter 5

Building Legendary Teams

The foundation of any transformative organization lies in its people. Great leaders recognize that their success is only as strong as the teams they build, nurture, and empower. In the fast paced, high stakes world of IT, legendary teams aren't just a nice to have, they are the very engine of innovation, resilience, and success. Building a legendary team is an intentional process. It requires more than assembling talented individuals; it demands creating a cohesive, dynamic group that shares a unified vision and the drive to achieve it.

In high-performing environments such as pharmaceutical R&D, aerospace engineering, and financial algorithmic trading, legendary teams are formed not just around talent, but shared mission fluency, high-feedback loops, and decentralized decision capacity.

Legendary teams have distinct energy. They exude collaboration, trust, and purpose. These aren't just teams that deliver results; they are teams that elevate everyone involved, setting a gold standard for what's possible. If you aspire to lead such a team, this chapter offers a roadmap to help you cultivate the elements that define legendary teams.

The Foundation of Legendary Teams: Vision and Alignment

Every great team begins with a compelling vision that acts as a guiding star. Without a shared purpose, even the most skilled group of individuals can become fragmented, struggling to find cohesion or direction. Vision provides clarity, motivation, and a sense of purpose that unites individuals and directs their efforts toward meaningful outcomes.

Do this: Run a **Vision Sprint** quarterly, ask team members to define success from three perspectives: customer, team, and mission. Capture these on a shared canvas to review monthly.

Use artifacts like team charters, quarterly value maps, or mission sprints to co-author vision and purpose, not just cascade it.

Your role as a leader is to articulate a vision that resonates deeply with your team. This isn't just about crafting a high-level mission statement, it's about defining a purpose that connects to both the organization's goals and the personal aspirations of the people you lead. A compelling vision answers questions like: **Why does this team exist? What difference will it make? Why should each member care?** The answers to these questions create a powerful sense of alignment and shared commitment.

In practice, run workshops using **north star canvases** to identify what success looks like from customers', operators', and systems' perspectives.

Alignment is essential to ensure that individual aspirations complement the broader goals of the team. Understanding what drives each person allows you to link their personal growth and contributions to the success of the group. When team members see how their unique skills and efforts contribute to a larger purpose, they become more engaged, motivated, and focused. The synergy created by this alignment amplifies the team's potential, transforming it into a unified force capable of extraordinary accomplishments.

Identifying and Attracting the Right Talent

The journey to building a legendary team begins with bringing the right individuals into the fold. Talent selection is not just about qualifications or technical skills; it's about identifying people with the mindset, adaptability, and passion necessary to thrive within your team's culture.

Adopt this: Design your hiring pipeline to assess for **team fit** early. Include behavioral prompts like: Tell me about a time you solved a problem with zero context and no clear owner. Watch how they handle ambiguity.

In regulated or high-risk domains such as defense, MedTech, or national infrastructure, talent frameworks should include values alignment, escalation behavior under stress, and mission adaptability as critical screening pillars.

Start by redefining what you look for in candidates. Instead of focusing solely on experience or certifications, prioritize qualities like curiosity, resilience, and a willingness to collaborate. A candidate's ability to learn, adapt, and bring fresh perspectives often outweighs their experience on paper. These are the people who will grow with your team, challenge the status quo, and drive innovation. Use situational judgment tests and rapid sprint assessments, two-hour problem-solving labs, to evaluate candidates' behavior under ambiguity.

Your recruitment process should reflect the culture you want to build. Go beyond traditional interviews and incorporate real world scenarios, collaborative exercises, or case studies. These methods provide valuable insights into how candidates think, communicate, and solve problems. Involving current team members in the hiring process, you gain a deeper understanding of how well a candidate might integrate and contribute to the existing dynamic.

Create reusable **culture in action** scorecards that measure how candidate responds to team norms like iteration, conflict resolution, or delivery accountability.

Creating a Culture of Trust and Collaboration

Trust is the bedrock of any successful team. Without it, even the most talented individuals can become disconnected and disengaged. Building trust requires consistent effort, transparency, and a commitment to fostering an environment where people feel safe to express themselves and take risks.

Adopt this: At the start of every new project or sprint, run a **Trust Reset** session, ask what assumptions people are holding, what worries them, and what they need from each other to succeed.

As a leader, you set the tone for trust within your team. Transparency in decision making, fairness in actions, and empathy in interactions demonstrate your commitment to creating a supportive environment. When team members see that you lead with integrity, they are more likely to mirror those behaviors and extend trust to their peers.

In industrial sectors where cross-functional teams operate under real-time constraints such as emergency response or aviation, trust is often operationalized through protocols like psychological safety checks at shift handoffs or daily mission debriefs.

Collaboration is the natural outcome of trust. It goes beyond simply working together, it involves leveraging the diverse strengths of each team member to achieve a common goal. Encourage open dialogue and active participation, ensuring that everyone's voice is heard and valued. Break down silos that hinder communication and promote cross functional projects that foster mutual understanding and cooperation. Celebrating collaborative successes, you reinforce the importance of teamwork and create a sense of collective achievement. Use pair rotations, shared OKRs, and role-switching days to build collaboration muscle and shared empathy between disciplines.

Empowering Individuals for Collective Success

Empowered individuals are the cornerstone of a legendary team. Empowerment is not about relinquishing control, it's about giving your team the confidence, autonomy, and resources they need to excel.

Handle this: Define three **ownership rings** for team decisions, Ring 1: Do it without asking. Ring 2: Propose it, then act. Ring 3: Escalate before proceeding. This builds guardrails for autonomy.

Build layered decision frameworks such as pre-approval zones for scope changes under **$X or autonomy** rings based on experience, to enable scalable ownership without creating chaos.

Begin by setting clear expectations for roles, responsibilities, and outcomes. Clarity provides a framework within which creativity can flourish, allowing team members to focus on delivering their best work. Provide the tools and training necessary to help them succeed and be available to offer guidance without micromanaging.

Investing in your team's growth is a powerful way to demonstrate your commitment to their success. Offer opportunities for professional development, whether through workshops, mentorship programs, or challenging assignments. Encourage them to take ownership of their learning and explore new areas of interest.

Acknowledging contributions is another critical aspect of empowerment. Celebrate achievements, both individual and collective, and show appreciation for the effort behind them. Recognition fosters a sense of pride and reinforces the behaviors that lead to success.

Institute impact recaps at the end of sprints or deliverables, short, public shoutouts that connect team wins back to mission outcomes.

Navigating Challenges with Resilience

Challenges are an inevitable part of any team's journey, but how you respond to them will define your team's resilience and character. Whether it's internal conflicts, external pressures, or unforeseen setbacks, legendary teams use adversity as an opportunity to grow stronger.

Do this: When conflict arises, ask two questions before any debate starts: **What are we trying to protect?** and **What can we safely test?** This reframes debate around value, not ego.

When conflicts arise, address them directly and constructively. Encourage open, respectful discussions that focus on solutions rather than assigning blame. Teach your team to view disagreements as a source of growth and innovation, rather than division. This approach fosters a culture where differing perspectives are welcomed and leveraged for better outcomes. Use structured models like **blameless postmortems** or **conflict ladders** to ensure that disagreement results in alignment, not fracture.

Setbacks should be reframed as learning opportunities. Share stories of challenges you've overcome, emphasizing the lessons learned and the resilience gained. This helps your team see failure as a natural part of growth rather than a permanent setback. Offer support and guidance to help them move forward with renewed confidence.

When external pressures mount, such as tight deadlines or high stakes projects, maintain a steady presence. Show your team that you believe in their ability to rise to the occasion and provide the resources and encouragement they need to succeed. Resilience is built through experience and a shared commitment to overcoming obstacles together. In sectors like logistics or critical infrastructure, run **crisis simulations** or **what-if** drills to train team response without waiting for real disruption.

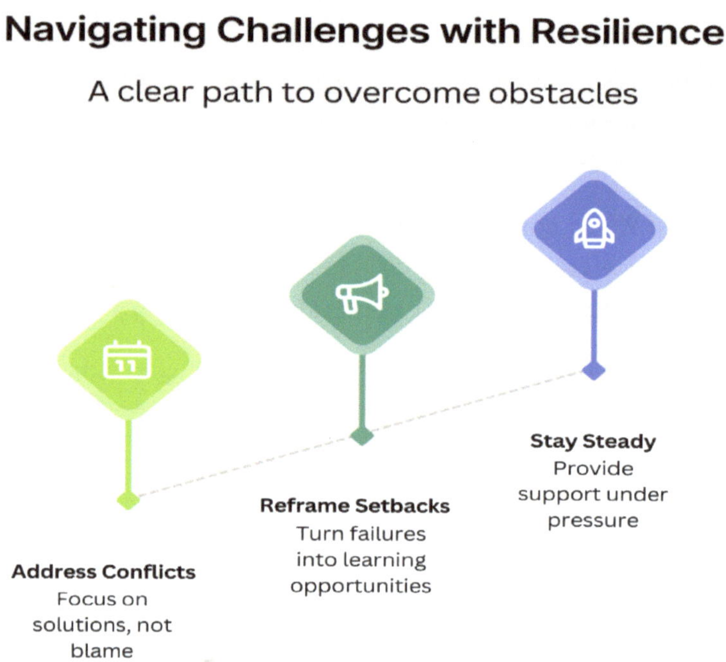

Fostering Innovation and Creativity

Legendary teams are not just performing high, they are trailblazing. They push boundaries, challenge norms, and deliver results that exceed expectations.

Adopt this: Introduce a **permission-to-experiment** budget, allocate 5-10% of team time for idea pilots with pre-defined sunset criteria. Reward experiments, not just results.

Encourage your team to think boldly and take risks. Innovation often involves exploring uncharted territory and embracing the possibility of failure. Diversity within your team is a powerful driver of creativity.

Build team rituals around innovation cycles such as 3x3 challenges where teams pitch three ideas every three weeks or allocate 10% of project time to unstructured R&D.

Measuring Success and Continuously Improving

Building a legendary team is not a one-time achievement, it's an ongoing journey.

Track this: Set up a **Team Health Index** with indicators like psychological safety, peer recognition, cross-team dependencies resolved, and delivery pride. Review this quarterly.

Establish clear metrics to evaluate performance, engagement, and impact. Solicit feedback frequently, both from within the team and from stakeholders. Continuous improvement requires a willingness to adapt.

Celebrate milestones and accomplishments to reinforce the culture you've worked to build. These moments of celebration inspire confidence and encourage your team to aim even higher.

Your Blueprint for Legendary Teams

As a leader, you have the power to shape a team that not only achieves success but also leaves a lasting legacy.

Build this: Create a **Legendary Team Manual** with principles, rituals, decision rights, and success stories. Let it evolve with the team.

This is your opportunity to create something truly remarkable. Begin building your legendary team today, and together, you can achieve greatness that transcends expectations and redefines what's possible.

Chapter 6

IT as a Catalyst for Business Growth

In today's dynamic business landscape, IT is far more than a support function, it is the engine that powers innovation and drives growth. IT has the potential to redefine how organizations operate, engage customers, and unlock new opportunities. However, for IT to act as a true catalyst for business growth, it must go beyond the implementation of tools and technologies. Success lies in the strategic integration of IT with organizational goals, the meaningful application of innovation, and a relentless focus on creating value across every facet of the enterprise.

In global sectors like retail, healthcare, and finance, IT has evolved into a value-generating function that enables competitive differentiation through digital-first business models. In these industries, IT teams are embedded in core product and service delivery, not just operational support.

This chapter delves into the transformative role of IT in driving business growth. It explores how IT leaders can align technology initiatives with corporate objectives, foster innovation, enhance customer experiences, and streamline operations to create an agile and resilient organization. Adopting these principles, IT can move from the background to the forefront, becoming a cornerstone of growth and competitive advantage.

Understanding the Intersection of IT and Business Growth

Business growth thrives on adaptability, innovation, and a deep understanding of customer needs areas where IT plays an indispensable role. IT serves as the backbone of operational efficiency, a driver of customer centric innovation, and a gateway to new markets. However, these opportunities can only be realized when IT strategies are fully aligned with the broader objectives of the business.

In logistics, for example, predictive route optimization and IoT-enabled fleet tracking are direct growth enablers that stem from IT-business alignment. In pharma, IT alignment fuels accelerated R&D through advanced data modeling.

IT AS A CATALYST FOR BUSINESS GROWTH

1. **Strategic Alignment**
   Integrate IT with organizational goals to drive value.
2. **Innovation Pipelines**
   Foster creativity, experimentation, and outcome-driven innovation.
3. **Customer Experience**
   Use IT to deliver personalization, convenience, and seamless interactions.
4. **Operational Efficiency**
   Streamline workflows, cut costs, and enhance productivity.
5. **Data & Leadership**
   Leverage data insights and empower IT leaders as growth partners.

Alignment is the critical factor that determines whether IT is a growth enabler or a bottleneck. Disconnected IT initiatives often lead to wasted resources, inconsistent customer experiences, and strategic misalignment. Every technological decision must be made with a clear understanding of how it contributes to overarching business goals. IT leaders must constantly evaluate whether their efforts are enhancing agility, improving customer satisfaction, and fostering innovation.

Implement this: Create a quarterly **IT-to-Growth** map where every tech initiative is explicitly linked to a business OKR. Review it with business unit leaders to adjust prioritization based on changing market signals.

Strategic Alignment: IT and Business Hand in Hand

For IT to drive growth effectively, it must function as an integral part of the business strategy rather than as a separate entity. This starts with ensuring that IT leadership has a voice in strategic decision-making.

In financial services, this means IT partnering directly with product teams to co-design secure, customer-facing tools that meet both compliance and innovation standards.

The alignment process begins with clear and open communication. IT leaders must actively engage with other business units to identify key objectives, whether it's entering new markets, increasing market share, or improving operational efficiency.

Use this tool: Launch **Joint Value Sprints** 2-week design cycles where IT, product, and business teams co-define use cases, prototype outcomes, and assign value targets before execution begins.

Operationalizing this alignment requires breaking down silos. IT teams should be embedded into cross-functional initiatives, ensuring their efforts are directly tied to business outcomes. For example, IT can work with marketing teams to develop data-driven campaigns or collaborate with operations teams to enhance supply chain management.

The best practice observed in manufacturing involves creating shared OKRs between IT, quality, and production teams to ensure that every technology investment improves yield, reduces downtime, or supports lean operations.

Building IT Inspired Innovation Pipelines

Innovation is often heralded as the key to growth, but in IT, it must be more than a buzzword, it must become a structured process that delivers tangible results.

Do this: Establish an "**Innovation Ops Model**" with three repeatable stages, Ideation, Incubation, and Integration, with exit criteria and funding tiers for each.

Establishing dedicated innovation sprints, funded pilot programs, or tech incubators within the IT organization are proven approaches used in sectors like MedTech and aerospace to systematically test, scale, and operationalize new ideas.

The foundation of innovation lies in empowering your team. Equip them with cutting-edge tools, provide them with the freedom to experiment, and encourage them to challenge conventional thinking.

Create a culture where failure is seen as an opportunity to learn rather than a setback.

Adopt **Safe-to-Fail Zones** for rapid innovation, bounded experiments that test hypotheses without requiring executive escalation.

Focus your innovative efforts on areas that align with your business objectives.

Tie each innovation program to a growth lever, revenue acceleration, customer retention, cost avoidance, or margin expansion, to ensure ongoing executive sponsorship and resource alignment.

Enhancing Customer Experiences Through IT

Customers are at the heart of business growth, and IT plays a pivotal role in creating exceptional customer experiences.

Apply this: Launch a **Customer Journey IT Overlay** where every touchpoint is mapped to a supporting IT capability, API latency, personalization rules, data syncs, etc.

Begin by gaining a deep understanding of your customers' needs and pain points. Use data analytics to uncover patterns and insights that can guide your strategies.

Invest in technologies that enhance customer engagement, such as mobile applications, chatbots, and self-service portals.

In consumer banking, cloud-native digital platforms now deliver frictionless onboarding, AI-powered financial advice, and secure multi-device experiences, each enabled by IT architecture designed for responsiveness and scale.

Integrating these tools with backend systems ensures a smooth and consistent experience across all touchpoints.

Operational Efficiency: Turning IT into a Growth Engine

Operational efficiency is a key driver of business growth, and IT is uniquely positioned to optimize workflows, reduce costs, and enhance productivity.

Try this: Build a **Digital Waste Heatmap** that surfaces process friction, human workarounds, and technology redundancy across business units.

In healthcare systems, implementing EHR automation, AI-based diagnostics, and robotic process automation has reduced patient cycle time while improving care coordination, translating directly into bottom-line performance.

Cloud computing, for instance, allows teams to collaborate seamlessly and scale resources as needed, reducing overhead costs. Similarly, data analytics tools can provide real-time insights that enable smarter decision-making and resource allocation.

Data as the Lifeblood of Growth

Data is one of the most valuable assets an organization possesses, and IT is responsible for unlocking its full potential.

Adopt this: Deploy a **Data-to-Growth Flywheel** model.

Capture → Analyze → Activate → Measure.

Assign business owners for each stage to ensure accountability.

In retail and CPG, real-time inventory analytics have enabled predictive restocking and demand forecasting, reducing stockouts by up to 30% and increasing revenue per square foot.

Establishing a robust data infrastructure is the first step. Implement systems that collect, store, and organize data in a way that makes it easily accessible and actionable.

Empowering IT Leadership to Drive Business Growth

For IT to act as a growth catalyst, its leaders must be empowered to make strategic decisions and drive change.

Create **IT Growth Portfolios** where leaders manage initiatives by business outcome, not just tech stack. Each portfolio has KPIs aligned to value delivered.

Support IT leaders with stretch roles such as interim general management of digital products or rotations through commercial teams to deepen business fluency.

Measure outcomes such as ROI, customer satisfaction, and operational efficiency to build trust and showcase IT's contribution to the organization's success.

Staying Agile in a Changing World

In industries like defense, energy, and telecom, adaptive IT systems built on microservices and DevSecOps have reduced deployment cycles from months to days.

Build this: Operationalize agility through **Release Cadence Dashboards** that track deploy frequency, incident rates, and feedback loop time.

Agility extends beyond technology; it must be ingrained in the mindset of your team.

Encourage an environment where experimentation, iterative development, and continuous improvement are celebrated as drivers of progress.

Calculated risk-taking is another cornerstone of agility.

Fostering this resilient and adaptive mindset, IT leaders can ensure that their teams remain forward-looking, nimble, and ready to seize new opportunities.

Creating a Culture of Collaboration

In large-scale enterprises, co-location models and fusion teams, where developers, analysts, and business leads work side-by-side on shared metrics, have accelerated cycle times and improved product-market fit.

Deploy internal collaboration frameworks such as **shared accountability maps** or **decision forums** to institutionalize cross-functional decision-making.

Try this: Set up a monthly **Tech-for-Business Lab** teams from IT and business units workshop one live use case together, producing a joint playbook in real time.

Regular cross-departmental meetings, joint workshops, and shared project ownership can break down barriers and build stronger relationships.

Leadership must model inclusivity, curiosity, and co-creation. When collaboration becomes an integral part of organizational culture, it fosters innovation, strengthens interdepartmental relationships, and enhances overall performance.

The Future of IT as a Growth Driver

As technology continues to evolve at an unprecedented pace, the role of IT in driving business growth is set to become even more profound.

Organizations that treat IT as a growth function are already staffing Chief Growth Technology Officers or embedding CIOs within M&A, venture, and innovation units.

One of the most significant opportunities lies in the adoption of AI-driven innovations.

Future-proof this: Assign a **Tech Horizon Lead** responsible for scouting, piloting, and de-risking disruptive technologies across each business domain.

Sustainability is another critical frontier where IT will play a pivotal role.

From optimizing data center operations to deploying intelligent IoT solutions that monitor and reduce resource consumption, IT leaders can ensure that their organizations remain ahead in the race toward sustainability.

The organizations that will thrive in the future are those that recognize IT as more than a technical function.

Aligning IT with business objectives, fostering innovation, and embracing an adaptive mindset, leaders can transform their organizations into agile, resilient powerhouses.

IT is not just a part of the business, it is the business's greatest opportunity to achieve sustainable growth and lasting success.

Chapter 7

Leadership During IT Crises

Leadership during a crisis is a defining test for any IT leader. It's the moment when the stakes are highest, and every action or inaction has far-reaching consequences. IT crises are not hypothetical scenarios; they are inevitable challenges that come in many forms, from cybersecurity breaches and system outages to failed implementations and unforeseen global disruptions. Such moments demand composure, rapid assessment, and decisive action backed by clear communication.

Strong leaders maintain visibility, reassure stakeholders, and mobilize their teams with confidence and purpose. The true measure of leadership lies in how effectively you respond, not just to resolve the crisis but to turn adversity into an opportunity to demonstrate strength, resolve, and competence.

This chapter explores the demands of leading through crises, offering strategies to prepare for the unexpected, act decisively under pressure, and emerge stronger. The objective is not merely to survive crises but to grow through them, ensuring your leadership becomes a beacon of trust and reliability.

Apply this mindset: View each crisis as a **trust multiplier** if managed with clarity and calm, crises can amplify your team's belief in leadership and accelerate cultural cohesion.

Embrace the Reality of Crisis Leadership

Crises are an inherent aspect of IT leadership. Systems fail, vulnerabilities are exploited, and unforeseen events disrupt carefully laid plans. The first step to mastering crisis leadership is to accept this reality and prepare for it with a proactive mindset. Rather than viewing crises as disruptions to be feared, they must be seen as challenges that test and refine your leadership abilities.

Preparation begins with cultivating a mindset that embraces resilience. A leader who remains composed, adaptable, and focused during a crisis instills confidence in their team and stakeholders.

Use a **Crisis Readiness Ledger,** a shared document tracking known fragilities across systems, vendors, and people. Review monthly in leadership forums.

In sectors such as healthcare and utilities, where system downtime can have life-or-death consequences, leaders conduct quarterly blackout drills or tabletop exercises simulating catastrophic events. These proactive simulations establish mental muscle memory that enables calm execution when real crises emerge.

Building a Crisis Ready Culture

The success of your crisis management efforts depends heavily on the culture you cultivate within your IT organization. A crisis ready culture does not develop overnight; it is the result of intentional leadership practices that prioritize preparedness, communication, and collaboration.

Encourage your team to anticipate potential risks and vulnerabilities. Simulations and drills, such as mock data breaches or system outages, help train your team to respond effectively under pressure. These exercises create clarity around roles and responsibilities, ensuring that everyone knows their part when the unexpected happens.

Codify this using **Crisis Role Cards** laminated or digital reference tools that assign decision scope, escalation path, and comms expectations by role.

Equally important is fostering a culture of transparency. Problems should be reported quickly and openly without fear of blame. When your team feels safe bringing issues to light, you gain critical time to address them before they escalate. This open communication also strengthens trust, which is invaluable during a crisis.

In public sector and defense environments, **failure blamelessness** policies are embedded in crisis response charters, ensuring that technical staff can escalate early without fear of retribution. As a result, threat containment windows are significantly reduced. Some leading organizations in manufacturing and finance use **Red Teams** internal groups tasked with attempting to break systems or expose vulnerabilities, to keep response teams agile and tested in live-fire exercises.

Implement **Signal Boost Protocols** auto-escalation triggers for certain incident types (e.g., unauthorized access, latency degradation) that alert designated war room teams via Slack, PagerDuty, or Teams.

Act Decisively, Even in Uncertainty

Crisis situations are often characterized by incomplete information and rapidly evolving circumstances. While uncertainty is unavoidable, hesitation can amplify the severity of the problem. The key to effective leadership is taking decisive action, even when all the details are not yet clear.

Start by gathering as much information as possible in the shortest time frame.

Use a **60-minute Triage Canvas** a template covering affected systems, customers impacted, legal exposure, and reputational risk. Used live during incident onset.

Field-proven playbooks often follow a 30-60-90 minute rule: contain immediate damage in the first 30 minutes, establish communication within 60, and stabilize key systems or issue customer updates by 90.

Communicate with your team promptly and clearly. Explain the immediate steps being taken and the reasoning behind them. A calm, decisive leader inspires confidence, encouraging the team to stay focused on resolution.

Communicating with Precision and Empathy

The importance of communication during a crisis cannot be overstated. Internally, your team needs clear, concise, and actionable instructions. Transparent about the gravity of the situation, but avoid causing panic.

Externally, transparency and accountability are crucial.

Deploy **Crisis Comms Kits** pre-approved email, press, and in-app message templates paired with ownership trees and sequencing guidelines for public releases.

Some organizations, especially in aviation and finance, maintain detailed chain-of-trust protocols. This ensures that messaging remains consistent and that no sensitive information is released prematurely or inaccurately.

Designate **Voice of Crisis** roles 1 external-facing lead, 1 internal translator (for execs/boards), and 1 incident scribe responsible for documenting decisions and timelines.

Learning on Your Crisis Management Plan

Crisis leadership is not about improvisation, it's about preparation. A robust crisis management plan serves as your blueprint for navigating the chaos. Assign specific roles and responsibilities to team members, ensure escalation paths are well defined, and test regularly.

Structure plans into 5 categories: Detection, Containment, Communication, Recovery, Reflection. Tie each to a named team.

If you don't have a plan, begin with common scenarios like cybersecurity breaches, system outages, and supply chain failures.

In regulated industries like banking, these plans are often tied to compliance audits and risk registers. Some embed automated alerts from SIEM platforms that notify execs the moment a severity threshold is breached.

Use **After-Action Packets** after simulations, executive brief templates summarizing what was learned, gaps revealed, and next actions with ownership.

Maintain Emotional Resilience

Leading through a crisis is as much about managing emotions as it is about managing systems. Emotional resilience is essential for navigating these high-stress situations effectively.

Your composure directly influences your team.

Deploy **Resilience Rotation Leads** non-technical roles assigned during major incidents to monitor fatigue, morale, hydration, and decision lag across teams.

Recognize stress in yourself and others. Techniques such as controlled breathing, reflection pauses, and even humor, when appropriate, can break tension and recenter focus.

Seek support when needed. Collaboration and input from others lead to better outcomes.

Use the **1-Hour Rule** every 60 minutes during extended incidents, pause for 5 minutes of debrief, hydration, and perspective reset. Burnout prevention is a leadership duty.

Turning the Crisis into a Learning Opportunity

Once the crisis is over, **don't move on, move forward**. Conduct a thorough review of the incident.

Schedule **Hot Wash** sessions within 72 hours, fast, candid, blameless discussions that prioritize speed of feedback capture over presentation polish.

Analyze root causes, assess your response, and document lessons learned. Sharing these insights reinforces a culture of readiness and learning.

Global companies mandate retrospectives that examine not just the technical failure, but also handoff timing, decision latency, and comms clarity.

Use **Crisis Replay Decks** slide summaries with incident timelines, decision inflection points, and stakeholder reactions, circulated in quarterly business reviews.

Build Resilience for the Future

Resilience is the ultimate goal of crisis leadership. Invest in monitoring systems, proactive alerts, robust infrastructure, and people readiness.

Maintain a **Resilience Register** a live tracker of systems, vendors, third-party risks, and mitigations. Assign a Resilience Steward for each function.

In sectors like pharma and utilities, crisis preparedness is written into vendor SLAs. Joint simulation tests and quarterly war room rehearsals are standard.

Also account for non-technical disruptions, workforce attrition, natural disasters, geopolitical volatility.

Integrate **Scenario Libraries** modular playbooks for disruption types beyond IT (e.g., reputational attacks, activist investor response, ransomware extortion).

Inspire Confidence and Trust

The way you lead during a crisis shapes your long-term credibility. Calm, decisive, and empathetic leadership inspires lasting trust.

Post-crisis, share **Leadership Reflections** brief internal notes or videos that outline what was learned, what will change, and how you see the team's response.

When managed skillfully, crises become inflection points. They expose fragility but also reveal character.

Repeat this mantra: **Every crisis is a capability reveal.** It surfaces what's working, what's missing, and who's ready.

Crises don't define your leadership by default, your response does.

Part 3

Metrics and Sustainability (Chapters 8–9)

Part 3 emphasizes the dual importance of measuring success and driving sustainability in IT leadership, guiding leaders to create meaningful impact while promoting responsible practices.

Chapter 8 focuses on metrics for legendary impact, offering practical insights into measuring the success of IT initiatives. Leaders learn how to track performance, demonstrate value to stakeholders, and align metrics with organizational objectives, ensuring every project contributes to broader goals.

Chapter 9 shifts attention to driving sustainability through IT, exploring the critical role technology plays in reducing environmental impact and fostering green practices. From energy efficient systems to reducing e waste, leaders are equipped with strategies to embed sustainability into their operations and align their initiatives with global environmental standards.

Together, these chapters highlight how IT leaders can balance operational excellence with ethical responsibility. Mastering metrics and championing sustainability, leaders not only ensure measurable success but also contribute to a more sustainable future for their organizations and the world.

Chapter 8

Metrics That Matter

Metrics are the lifeblood of an IT organization, providing insight into operations, signaling successes, and serving as early warnings for potential challenges. Without metrics, decision making becomes guesswork. With the right metrics, you unlock clarity, drive transformation, and showcase IT's role as a strategic force within your organization. Consider metrics not only as diagnostics but as leadership currency, used to align priorities, influence funding, and build executive trust..

To lead effectively, you must adopt a rigorous and purposeful approach to metrics. They are your compass, your guide to where you stand and where you need to go. Yet, not all metrics are created equal. Some are distractions, while others are invaluable signals. Your role is to cut through the noise, focus on what truly drives value, and use metrics as a bridge between IT initiatives and business objectives.

Build a **Metrics Ladder** a 3-tier model connecting team-level KPIs, operational performance, and strategic business outcomes to reduce misalignment.

selecting and refining the right metrics, leaders can create a shared language across departments, ensuring IT is not viewed as isolation but as an integrated partner in strategy. Metrics become not just tools for monitoring, but instruments of influence that **inspire accountability, shape culture, and reinforce** the organization's long-term vision.

Understanding the Power of Metrics

Metrics go beyond numbers on a dashboard; they tell the story of your IT organization's health, efficiency, and impact. They illuminate opportunities for growth, uncover vulnerabilities, and provide the evidence needed to validate decisions.

Use a **Metrics-to-Impact Map** a visual linking each IT initiative to its downstream business effect, e.g., cost saved, risk reduced, satisfaction gained.

A well-crafted metrics framework articulates IT's value, strengthens stakeholder trust, and positions IT as a strategic enabler of business objectives.

The power of metrics lies in their alignment with organizational goals. Metrics disconnected from the broader business strategy risk becoming irrelevant or counterproductive. For instance, tracking server uptime without understanding its impact on customer satisfaction or revenue growth reduces its strategic significance.

Apply **Signal-Purpose-Outcome** framing: every metric must have a defined signal (what it detects), a purpose (why it matters), and an outcome (what action it influences).

Metrics also empower storytelling. They allow IT leaders to communicate complex ideas in tangible terms.

In regulated sectors like banking and government, stakeholders often require quantitative proof of value in boardroom reporting.

Healthcare systems, for example, now track EMR latency and digital triage speed as operational quality indicators, not just IT outputs.

Adopt **Executive Metric Cards** single-page visuals used in board decks that define each metric's owner, narrative, and outcome pathway.

Crafting a Metrics Strategy

Effective metrics don't happen by accident; they require a deliberate strategy. Begin by identifying the outcomes you aim to achieve.

A disciplined metrics strategy avoids the temptation to measure everything.

Limit dashboards to a **Top 10** ruleset, no more than ten primary metrics per domain to maintain executive focus.

Instead, focus on metrics that are:

- ✓ **Relevant:** Directly tied to organizational and IT objectives.
- ✓ **Actionable:** Providing clear insights that guide decision making.
- ✓ **Timely:** Delivering real time or near real time information when it's most useful.
- ✓ **Transparent:** Clear and understandable for all stakeholders, fostering trust.

Think of your metrics framework as a living entity.

Embed **Quarterly Metrics Pulse** rituals cross-functional forums to retire stale metrics, nominate new ones, and align visibility with changing priorities.

High-performing IT organizations often create metrics councils. These include finance, HR, and customer service to ensure IT metrics resonate beyond the tech stack.

In manufacturing, deploy **Digital Downtime Dashboards** to correlate IT outages with production impact per plant or line.

In retail, couple uptime metrics with POS throughput velocity to demonstrate how digital reliability drives margin-per-foot.

Crafting a Metrics Strategy

A DELIBERATE FRAMEWORK TO DEFINE, REFINE, AND ALIGN METRICS WITH ORGANIZATIONAL GOALS.

Metrics should evolve with changing priorities, ensuring continuous alignment with organizational success.

The Three Dimensions of Metrics

Metrics fall into three primary dimensions: Operational, Strategic, and Predictive. Together, they create a full picture of IT value.

Operational Metrics

Operational metrics focus on the day-to-day functioning of IT systems. They measure reliability, efficiency, and performance.

Examples include uptime, MTTR, ticket volumes, and SLA compliance.

Add **Service Experience Indexes** hybrid scores combining response time, resolution accuracy, and satisfaction per incident type.

In logistics and utilities, operational metrics are used to track digital readiness of remote depots, not just headquarters.

Use **Field Reliability Panels** fleet dashboards visualizing mobile worker enablement (device uptime, app latency, signal failure rate).

Strategic Metrics

Strategic metrics connect IT to business value.

Examples: ROI on tech investments, digital NPS, or digital adoption rates.

Introduce **Adoption Heatmaps** visual grids tracking department-by-department usage of new tools, platforms, or services post-launch.

In financial services, cost-to-income ratios now include IT productivity.

Public agencies monitor how many residents' complete services digitally to gauge progress toward digital government goals.

Use **Self-Service Share** metrics percentage of transactions completed via digital channels without human intervention.

Predictive Metrics

Predictive metrics help you anticipate what's coming.

Examples: anomaly detection, user churn risk, or system failure trendlines.

Build **Proactive Friction Models** combining latency, login errors, or usage drop-offs to surface silent UX issues before they trigger escalations.

In finance and e-commerce, these metrics help IT leaders prevent failure, not just detect it.

In energy, IoT-based predictive maintenance has blurred OT and IT.

Extend your dashboards to include asset degradation forecasts from plant SCADA data or remote sensor drift.

The Three Dimensions of Metrics

A holistic view of IT performance across operational, strategic, and predictive lenses.

Operational Metrics

Measure reliability, efficiency, and performance (e.g., uptime, MTTR)

Strategic Metrics

Link IT initiatives to business outcomes (e.g., ROI, NPS)

Predictive Metrics

Use data trends to anticipate challenges and opportunities.

Together, these dimensions ensure IT drives stability, business value, and future readiness.

Measuring What Matters

Start by aligning your metrics with the outcomes that define success. If you care about customer satisfaction, measure performance at the experience layer, not just uptime.

Create **North Star Bundles** linked metric groups that track a shared goal (e.g., for CX: latency + satisfaction + resolution time + retry rate).

Ambiguity is the enemy. Every metric needs:

- A clear definition
- A target range
- A timeframe for improvement
- An owner responsible for action

Present metrics in a digestible, visual way.

Use **Executive Action Tiles** dashboard cards showing a red-yellow-green status, last change, and owner call-to-action.

Avoiding Metric Pitfalls

The value of metrics lies in their quality, not their quantity. Common pitfalls include:

- ✓ **Data Overload:** Tracking too many metrics dilutes focus and creates confusion.
- ✓ **Lagging Indicators:** Solely relying on past performance metrics limits proactive decision making.
- ✓ **Misaligned Goals:** Metrics that don't reflect organizational priorities undermine credibility.
- ✓ **Lack of Ownership:** Every metric should have a designated owner responsible for its monitoring and outcomes.

Avoiding these pitfalls ensures your metrics become a consistently reliable tool for fostering accountability and driving meaningful, ongoing improvement throughout your organization.

Run a **Metrics Maturity Audit** a quarterly review checklist to score each metric's actionability, clarity, alignment, and adoption.

Using Metrics to Drive Accountability

Metrics are a backbone for accountability.

Make metrics personal: show how team-level KPIs contribute to division or company goals.

Example: Link ticket resolution speed to individual recognitions.

In leadership roles, your ability to speak metrics fluently earns trust and expands influence.

Incorporate **Metrics Check-Ins** during weekly ops reviews, each owner presents their key metric shift, what caused it, and their next move.

Celebrating Success and Learning from Data

Metrics can inspire when they show progress.

Use **Metric Wins Boards** as a digital space where teams pin and narrate improvements (e.g., MTTR dropped by 28% after alert routing fix).

Equally, failures in metrics are signal, use them to adjust systems, not assign blame.

Create a **From Red to Green** review a space where teams share how they turned around a failing metric with process or tech changes.

Your Metrics Journey

Metrics are more than numbers, they are the operating system of modern leadership.

Your dashboard is your cockpit, treat it as the live interface between IT operations and strategic growth.

Build a metrics framework that adapts, connects teams, and drives action.

Mastering metrics means learning to see beyond data, to insight, influence, and impact.

The tools are in your hands. Use them to tell a story of transformation, [one signal at a time.]

Chapter 9

Staying Ahead of the Curve

The pace of technological advancement is accelerating, demanding IT leaders rise above the noise and define the direction for their organizations. This isn't about simply keeping up, it's about stepping ahead, navigating with confidence, and ensuring that your vision shapes the future instead of reacting to it.

Staying ahead means more than adopting the latest tools or processes; it requires a mindset deeply rooted in curiosity, adaptability, and purposeful action.

It also demands structured foresight systems, embedded experimentation loops, and institutional memory, so that innovation becomes discipline, not chance.

As an IT leader, your role extends beyond implementing solutions, you are the architect of innovation, resilience, and progress.

To lead effectively in a world of constant change, you must foster a culture that not only embraces transformation but thrives on it.

Frame this by integrating **anticipatory execution** into leadership routines, don't wait for directives; prototype, test, and signal where change should go.

This means empowering your teams to experiment, learn, and grow while maintaining a clear focus on long term goals.

This chapter provides the roadmap to achieving that edge.

Use it to establish a repeatable framework, trend monitoring, readiness mapping, capability rotation, and bold delivery.

Together, we'll uncover the strategies that position you not just as a participant in the fast-moving world of IT but as a leader who sets the pace and inspires others to follow.

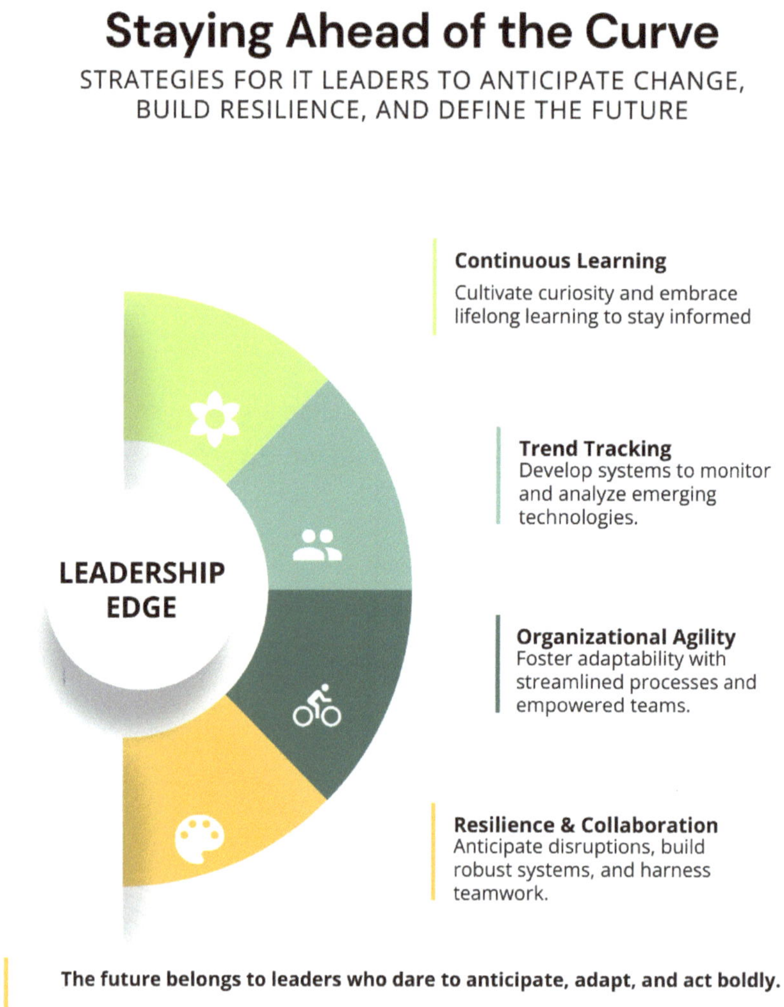

Embrace Continuous Learning and Curiosity

To lead effectively in an ever-changing landscape, you must cultivate a mindset of curiosity and a commitment to lifelong learning. The most successful IT leaders are those who continually seek to understand emerging technologies, trends, and methodologies.

Set Learning OKRs objectives that track both personal and team-level upskilling across quarters, tied to strategic initiatives.

Dedicate time regularly to explore areas beyond your immediate expertise. Engage with industry publications, attend conferences, and participate in webinars to remain informed about the latest advancements.

Use a **Learning Hours Ledger,** a simple shared log tracking how many hours each team member dedicates monthly to strategic learning.

While it's important to stay current with developments in your industry, broadening your scope to adjacent fields often provides unexpected insights.

For example, studying advancements in environmental technology could inspire innovative approaches to data sustainability within IT.

Encourage a culture of learning within your team. Provide access to professional development resources, sponsor training programs, and create opportunities for collaborative learning.

Operationalize this with **Peer-Led Teachbacks,** have team members present learnings from external sources monthly, building internal teaching capacity.

Adopt a recurring learning cycle. In healthcare IT, monthly clinical tech briefings keep cross-functional teams updated on both medical and IT innovations. You can replicate this cadence in your sector, rotating spotlight topics from AI to compliance with customer experience.

Institutionalize **Innovation Rounds** a recurring team-led forum where new tools, pilot results, or industry shifts are shared and debated.

Keep the flow constant and collective.

In government and defense, agencies now embed tech futurists in strategy roles.

Consider appointing a **Readiness Scout** a rotating team role focused on identifying horizon signals and translating them into capability gaps or opportunities.

Develop a System for Tracking Trends

The key to staying ahead is having a structured approach to monitoring trends.

Implement **Trend Intelligence Pipelines** structured roles, tools, and rituals that regularly collect, filter, and translate insights into roadmaps. While it's easy to be overwhelmed by the sheer volume of information available, a disciplined system ensures that you capture the most relevant and actionable insights.

Start with trusted sources leading publications, reputable analysts, and reliable industry forums.

Assign team members specific areas of focus, such as artificial intelligence, cybersecurity, or digital transformation, and have them regularly report their findings.

Expand your view beyond your immediate industry.

Use a **Cross-Pollination Matrix** compares signals across at least three adjacent industries (e.g., mobility, media, infrastructure) to identify transferable innovations.

Trend tracking is not merely observational; it's a proactive exercise.

Designate **Horizon Review Blocks** during quarterly planning short, tactical sessions where horizon scans inform backlog reshaping.

In manufacturing IT, trend data from wearables and predictive sensors are now reviewed alongside product roadmaps.

Use a **Time Horizon Table** to classify each signal as near-term (6–12 months), mid-term (12–24 months), or long-term (3–5 years). Revisit the table quarterly.

Foster Organizational Agility

Agility is the hallmark of organizations that not only survive but thrive in periods of rapid change. Design and test an **Agility Model** for your org structure, governance, tooling, and decision velocity benchmarks, iterated quarterly. Building agility begins with streamlining processes and removing unnecessary bureaucracy.

Empower teams at all levels to make decisions.

Reduce friction using a **2-Tier Approval Model** shifting routine operational decisions to the team level, retaining only high-risk escalations centrally. Agility also involves embracing iterative processes like Agile or Scrum, which allow for flexibility, regular reassessment, and continuous alignment.

Invest in technology that supports rapid adaptation.

Score every core platform quarterly on adaptability metrics, deployment speed, user configurability, failure containment, and scaling friction.

Adopt this mindset in execution: Build internal tiger teams capable of rapid deployment.

Assemble your team of cross-functional high performers now, before disruption demands it.

Create a **Readiness Roster** a pre-assigned list of leaders, analysts, engineers, and comms specialists who can be activated for emergent priorities.

Model this structure in pilot programs before scaling it enterprise-wide.

Anticipate Disruption Before It Happens

One of the most critical aspects of staying ahead is the ability to anticipate and prepare for disruption.

Adopt a proactive mindset by regularly asking "**what if**" scenarios.

Use a **Strategic Foresight Tabletop** quarterly role-play simulations of high-impact, low-frequency events such as regulatory shocks or mass system failures.

Leverage predictive analytics and emerging technologies to forecast potential disruptions.

Execute quarterly red-team simulations.

These exercises often surface gaps in decision authority, communication flow, or digital resilience.

Codify findings into an **Execution Delta Map** documenting where latency, ownership, or blind spots appeared, then directly tying them to sprint-level fixes.

Adopt disruption readiness as a budget line.

Establish a **Strategic Agility Reserve** a flexible, pre-approved funding block allocated solely for mid-year pivots, moonshot ideas, or shock absorbers.

Build Resilience in Your Strategies

Resilience is the ability to withstand and recover from disruptions.

Start by ensuring your infrastructure and business processes are designed to handle the unexpected.

Create a **Continuity Capability Ledger** listing essential IT, process, and human elements required to maintain 30/60/90-day viability.

Implement robust cybersecurity, reliable backups, and disaster recovery plans.

Regularly test these via **Dual Failure Drills** where two simultaneous failures (e.g., power + ransomware) are tested to reveal interdependencies.

Cultivate resilience within your team by encouraging a growth mindset.

Lead with transparency and optimism during challenging times.

Use structured post-mortem reviews.

Hold **Lessons Unlocked** sprints, each one closing with a documented fix, reusable template, or automation to prevent recurrence.

Harness the Power of Collaboration

Collaboration is an essential driver of innovation and progress.

Break down silos and encourage cross-functional teamwork.

Create **Problem Fusion Labs,** temporary squads of ops, IT, and business leads co-designing solutions to shared challenges.

Externally, seek partnerships with startups, research institutions, and other organizations.

Engaging with customers is equally critical.

Embed **Customer Feedback Loops** into product sprints, translating observations directly into iteration backlogs.

Lead collaboration by co-designing.

Replicate embedded prototype labs, digital sandboxes where frontline employees and IT co-build experiences.

Adopt external collaborations through innovation alliances.

Use **Consortium Maps,** a visual inventory of joint ventures, think tanks, or standards groups your org engages with, ensuring strategic coverage.

Commit to Bold Action

Leadership requires decisiveness, especially in rapidly changing environments.

Taking bold action involves balancing calculated risks with informed decisions.

Encourage a culture of experimentation where failure is part of the process.

Define a **Failure Operating Range** pre-approved boundaries within which teams can take action without escalating for permission.

When opportunities arise, act with clarity and purpose.

Build your execution reflex by launching internal accelerators.

Run 90-Day Innovation Sprints, dedicated time-boxed efforts with small, focused teams aiming to deliver measurable breakthroughs.

Handle risk by codifying your threshold.

Document your **10x Tolerance Window** how much volatility is acceptable for how much potential upside and socialize it to unblock decisions.

The Path Forward

Staying ahead of the curve is a continuous process that requires vigilance, adaptability, and determination.

It's about creating an environment where curiosity, resilience, and collaboration thrive.

Embracing learning, anticipating disruption, and acting boldly, you can lead your organization to not only keep pace with change but to define it.

Let these principles drive quarterly action, linking your leadership rhythm to execution rituals that prove strategic foresight.

The future belongs to those who dare to shape it. As a leader, you have the tools and the influence to stay ahead.

Embed them, use them, scale them.

Let this chapter be your guide as you navigate the complexities of a dynamic world, turning challenges into opportunities and possibilities into realities. The journey begins now.

Part 4

Staying Ahead and Adapting (Chapters 10–13)

Part 4 delves into the strategies and mindsets IT leaders need to thrive in a dynamic, ever-changing landscape. **Chapter 10**, Staying Ahead in the IT Landscape, provides actionable insights into adapting to emerging trends, technologies, and shifting market demands, ensuring leaders remain competitive and thinking forward.

In Chapter 11, The Legendary IT Mindset, the focus shifts to the philosophies and behaviors that define exceptional leaders. This chapter highlights the importance of vision, adaptability, and resilience in navigating challenges and seizing opportunities.

Chapter 12, Future Proofing IT Leadership, prepares organizations for long term success by emphasizing resilience and strategic planning. Leaders learn to anticipate changes, build adaptable systems, and guide their teams toward sustainable growth.

Finally, Chapter 13, The Global Visionary IT Leader, explores the unique challenges and opportunities of leading across borders, cultures, and industries. This chapter equips leaders with the tools to embrace diversity, foster collaboration, and drive global innovation.

Together, these chapters empower IT leaders to stay ahead, adapt, and lead with confidence in a constantly evolving world.

Chapter 10

The Legendary IT Mindset

The IT landscape is dynamic, marked by rapid advancements and constant disruptions. Yet, the true mark of a leader isn't merely technical expertise; it's the mindset that drives innovation, resilience, and purpose. A legendary IT mindset isn't innate, it's cultivated. It's about embracing challenges as opportunities, transforming setbacks into lessons, and crafting a vision that motivates teams and shapes the future. This chapter explores the essential elements of a legendary mindset and how to nurture it, ensuring you thrive as an IT leader. **This is your operating system. Mindset isn't abstract, it's a deployable system you design, upgrade, and scale**.

Why Mindset Defines Leadership

Mindset shapes everything, from how you react in moments of crisis to how you inspire those around you. While technical knowledge and strategic planning are vital, they mean little without the mental framework to adapt, lead, and innovate under pressure. A strong mindset is what allows you to turn uncertainty into clarity, setbacks into progress, and potential into legacy.

Imagine navigating a major outage affecting critical business operations. Technical tools can diagnose the problem, but the confidence to lead under pressure, the ability to communicate effectively, and the foresight to minimize future risks come from your mindset. It's the foundation that underpins not just your actions but how others perceive your leadership.

Adopt the mindset daily by anchoring your routines around outcome ownership. Begin your week by choosing one friction point, technical, cultural, or operational, where your influence will unstick progress. End the week by reflecting: What shifted because you showed up with intent? Build momentum by repeating this ritual.

Mastering Visionary Thinking

Visionary thinking is at the heart of a legendary mindset. It's not enough to focus on immediate tasks; you must think expansively, connecting present actions to long term outcomes. **Leaders with vision see patterns where others see chaos and anticipate opportunities where others see obstacles.**

Cultivating this skill starts with making time for reflection. Dedicate regular intervals to analyze trends, explore emerging technologies, and evaluate how these elements align with your organization's future. Visionary thinking requires an openness to explore ideas that might seem distant or speculative. For example, understanding how quantum computing could influence supply chain optimization might seem unnecessary today, but it could be pivotal five years from now.

Equally important is communicating your vision effectively. Break it down into actionable steps that your teams can embrace and execute. Visionary thinking isn't a solitary act; it's a collaborative journey where your ability to inspire others turns abstract ideas into concrete achievements.

Lead this by codifying your vision into a three-tier system: Now (what must evolve immediately), Next (12–18 months strategic positioning), and Beyond (emergent opportunities worth exploring). Use this framework consistently, in town halls, sprint reviews, and leadership 1:1s. Let your clarity become cultural currency.

Cultivating Adaptability in the Face of Change

Change is the norm in IT, and those who resist it risk obsolescence. Adaptability is more than just reacting to new circumstances; it's about thriving amid uncertainty and leveraging change as a strategic advantage.

Start by cultivating a mindset that welcomes experimentation. Encourage your teams to take calculated risks and innovate without fear of failure. Each misstep becomes a steppingstone to refinement and success. When a deployment falters or a solution underperforms, reframe these as opportunities to improve.

Flexibility is another critical component. Leaders must be willing to pivot strategies or abandon outdated processes when circumstances demand it. This doesn't mean abandoning structure but creating processes that can adjust to changing conditions. Whether adopting new cloud technologies or redesigning workflows, adaptability ensures your leadership remains relevant.

Institutionalize adaptability by holding "**Kill Your Darlings**" retrospectives quarterly. Challenge your team to question what tools, processes, or priorities no longer serve the mission. Celebrate not only what you build, but what you courageously dismantle.

Building Confidence Through Self Awareness

True confidence is grounded in preparation, self-awareness, and humility. It's not about projecting certainty in every decision but about demonstrating competence and earning trust.

Building confidence starts with preparation. Stay informed about technological advancements, industry trends, and organizational goals. Equip yourself with knowledge so that when challenges arise, you can approach them with assurance.

Self-awareness complements preparation. Regularly reflect on your strengths and areas for improvement. Seek constructive feedback from peers and team members to uncover blind spots. Confidence doesn't mean you always have the answers, it means you're willing to learn, adapt, and grow.

Practice reflective leadership by keeping a decision log. Capture your logic behind big choices, revisit them later, and ask: **What did I assume? What surprised me? What would I repeat?** Over time, this becomes your personal playbook of pattern recognition and growth.

Leading with Emotional Intelligence

Great IT leaders are not only technically proficient but emotionally intelligent. Emotional intelligence, understanding and managing emotions, both on your own and others, is the bridge between technical skill and inspirational leadership.

Self-regulation is the cornerstone of emotional intelligence. **When faced with high pressure situations, take a moment to pause and assess. Responding with clarity and thoughtfulness, rather than reacting impulsively, strengthens your credibility.**

Empathy is equally vital. Make time to understand the challenges and aspirations of your team members. Whether navigating a team conflict or supporting an overburdened employee, empathy creates an environment where people feel valued and supported. Such a culture fosters trust, collaboration, and innovation.

Finally, emotional intelligence requires resilience. IT leaders often face criticism or resistance, but those who remain composed and solution focused inspire loyalty and confidence.

Model this by leading from the center, not the edges. When tensions rise, slow the tempo. Speak in measured rhythms. Use silence strategically. Build emotional radar through informal skip-levels and hallway conversations, not just survey data. Let insight shape your tone.

Harnessing Curiosity to Drive Innovation

Curiosity is a defining trait of legendary IT leaders. It fuels creativity, exploration, and the ability to uncover groundbreaking solutions. Leaders who ask, **"What if?" and "Why not?"** open doors to possibilities that others might overlook.

To cultivate curiosity, engage in continuous learning beyond the confines of IT. Explore fields like behavioral science, economics, or design. Cross disciplinary insights often lead to transformative ideas. For instance, lessons from behavioral economics might inform user friendly IT system designs.

Encourage your team to adopt a curious mindset by fostering an environment where questions are welcomed and exploration is rewarded. Regular brainstorming sessions or innovation challenges can ignite fresh ideas and expand horizons.

Institutionalize curiosity by setting monthly learning provocations. Assign team members a **discovery sprint** 20 minutes of exploration in a new domain. Then, in a shared forum, they answer: **What surprised me? What could we try?** Curiosity scales when it's shared.

Inspiring Discipline and Drive

Discipline and drive are the engines of consistent success. Driving fuels your ambition to push through challenges, while discipline ensures that your efforts are focused and effective. Together, they transform potential into achievement.

Reconnect regularly with your purpose as an IT leader. Reflect on what excites you about solving complex problems and leading innovation. This sense of purpose will sustain your drive, even during tough times.

Discipline, on the other hand, is about structure and habits. Prioritize your tasks based on impact, avoid distractions, and set routines that enhance productivity. For example, scheduling uninterrupted time for strategic thinking or allocating periods for mentorship can significantly boost your effectiveness.

Run your calendar like a mission tracker. Every Friday, audit your time: **How much went to fire drills versus future shaping?** Shift your energy accordingly. Track three habits: one strategic task completed, one teammate empowered, one risk surfaced. Small daily wins drive disproportionate momentum.

Transforming Potential into Legacy

A legendary IT mindset transcends personal achievement. It's about creating a lasting impact that shapes your organization and the people within it. This requires a commitment to continuous growth, the humility to learn from others, and the vision to inspire future leaders.

Legacy isn't built overnight. It's the culmination of consistent actions, thoughtful decisions, and a willingness to adapt and evolve. Embracing the principles outlined in this chapter, you can create a mindset that not only drives success today but also leaves a lasting imprint on the IT landscape.

The legendary IT mindset is your greatest asset. It empowers you to navigate uncertainty, inspire your teams, and turn challenges into opportunities. Cultivate it, embody it, and let it guide your journey as a transformative leader.

Document your leadership fingerprint, your non-negotiables, the decisions you want others to understand, the practices worth passing down. Share them through mentorship, not memos. Identify two emerging leaders every quarter to guide, challenge, and equip. Your influence scales when others internalize and evolve your mindset without you in the room.

The Legendary IT Mindset

Harnessing curiosity, discipline, and vision to leave a legacy.

Chapter 11

The Ethics of IT Leadership

The ethical dimension of IT leadership has never carried more weight. Decisions about automation, data, or responding to security risks must balance progress with principle, safeguarding individuals as well as society. Ethics also functions as a safeguard for innovation. Without clear guardrails, new technologies risk eroding the very trust they aim to build. Leaders who prioritize transparency, accountability, and fairness can strengthen organizational credibility and maintain public confidence.

This chapter introduces strategies and mindsets for leading with integrity, proving that ethical leadership fuels both organizational performance and the lasting trust required for meaningful innovation. Begin by treating ethics not as a passive set of constraints but as an active leadership advantage. Approach every strategic decision through a lens of impact: **Who benefits? Who bears the cost? How can technology serve people, not just process?** Your ethical discipline becomes a differentiator.

Embed this mindset operationally by institutionalizing a decision checkpoint in every IT governance workflow labeled "Ethical Consequences." Require justification across four dimensions, user rights, long-tail risks, system equity, and transparency, and make this discussion a visible gate in project reviews.

Understanding the Stakes: Why Ethics Matters in IT Leadership

In IT, the stakes of ethical leadership are uniquely high. The systems you design and the decisions you make have ripple effects far beyond your organization, influencing customer trust, employee welfare, and even societal norms. Technology is powerful but neutral, it can uplift or exploit, depending on the hands guiding it.

Ethical leadership starts with acknowledging this responsibility. When your organization develops an AI model, implements new cybersecurity measures, or collects user data, the key questions must be: **Are we protecting those who rely on us? Are we aligned with our values? Will we look back on this decision with pride?**

This commitment to ethics isn't about avoiding scandal or fines. It is about fostering enduring trust, the kind of trust that retains loyal customers, attracts top talent, and reassures stakeholders that your organization acts responsibly. Ethical practices shape reputation, influencing how regulators, investors, and partners view your organization. A strong ethical foundation creates opportunities for collaboration, secures market credibility, and builds resilience against evolving risks.

True ethical leadership strengthens resilience during crises, ensuring decisions are made with long-term impact in mind rather than short-term gain. Leaders who ground their strategies in fairness and accountability send a signal that technology can be both transformative and humane. In an era of skepticism and scrutiny, ethical behavior becomes a competitive advantage.

Operationalize this advantage by requiring every project charter to include an **Ethical Risk Statement** reviewed by a multidisciplinary panel. Rotate ethics stewards across projects so that responsibility is shared, not siloed.

Ethical IT leadership is built on **six principles**: transparency through honesty, accountability in outcomes, equity ensuring fairness, privacy and security protecting data, sustainability for long-term impact, and integrity guiding values. These principles shape trust, strengthen resilience, and ensure technology serves people responsibly.

Link each principle to measurable behavior.

Example: **Transparency** = publish rationale for algorithmic changes, **Equity** = document bias-testing results for user impact.

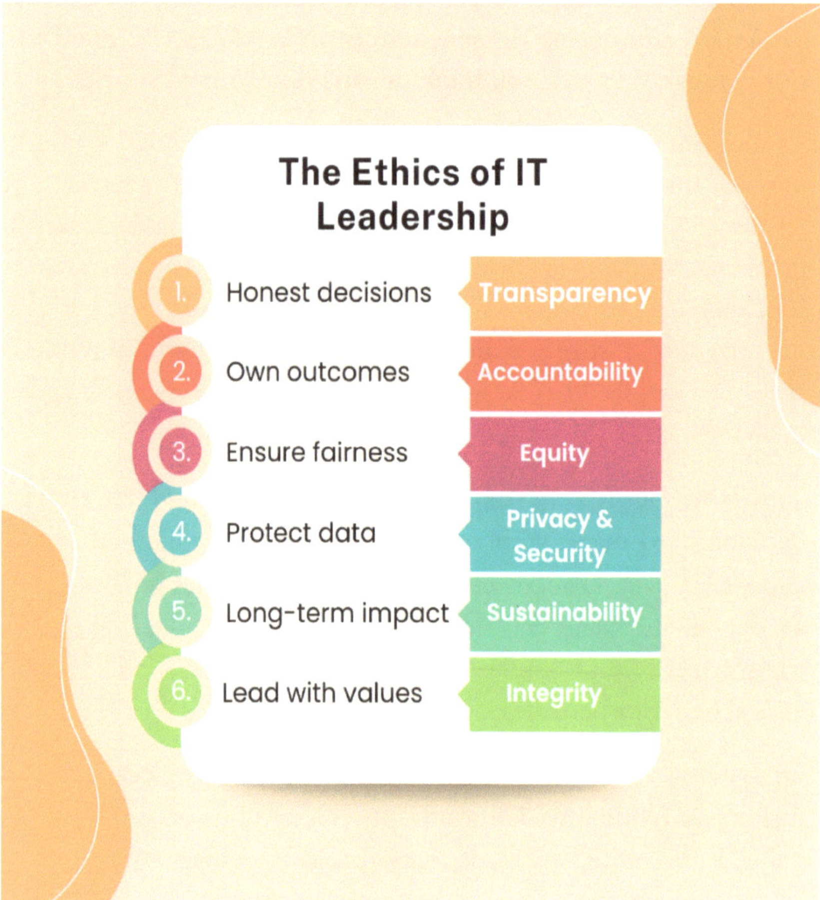

Applying Ethical Frameworks in Decision-Making

Ethical dilemmas in IT rarely have simple answers. Leaders must move beyond instinct and bias by applying structured frameworks. Two of the most influential, Kant's deontological ethics and utilitarianism, offer complementary perspectives when stakes are high and consequences far-reaching. Together they provide clarity, helping decisions align with long-term values and organizational trust.

Integrate these frameworks into your leadership review process. During solution approvals, force deliberate pauses to ask: **Does this respect dignity? Does it minimize harm? Is this a decision I could defend publicly?** Add an **Ethical Design Review** beside performance, scalability, and security reviews.

Kant's Deontological Ethics

Kant's system is rooted in duty and universal moral principles rather than consequences. The categorical imperative demands that every action be universalizable, if everyone acted this way, it must still be morally acceptable, and that people always be treated as ends in themselves, never merely as means.

Applied to IT leadership, Kant's framework demands that decisions respect human dignity, fairness, and autonomy. For example, when designing AI systems, leaders should prioritize transparency and fairness even if it slows down deployment or reduces profit. Collecting user data without consent may generate short-term business value, but under deontological reasoning, it fails the moral test because it disregards the inherent rights of individuals. A Kantian leader in IT therefore focuses on principles, privacy, honesty, accountability, ensuring they are upheld consistently across all decisions.

Implement **Consent Impact Reviews** for all user data systems. If consent is ambiguous, defer launch. Require design teams to present a Human Impact Statement before executive signoff.

Use **stakeholder mapping** tools like RACI but upgraded to include ethical exposure for each user group. Create a dashboard that logs risk acceptability thresholds by audience, and updates with every release.

Utilitarianism

Utilitarianism evaluates morality through outcomes, aiming to maximize overall well-being and minimize harm, the greatest good for the greatest number. It is particularly relevant in large-scale IT operations where millions of users may be affected.

For example, rolling out a new cybersecurity system may inconvenience employees but significantly reduce risk and protect customer data. The short-term cost is outweighed by long-term safety. Likewise, automation may displace some roles yet improve efficiency and create broader societal benefits.

Model these trade-offs with a **Net Benefit Matrix** during planning. Use axes for short-term pain, long-term value, and user-group scale, documenting risk-mitigation measures before approval.

Virtue Ethics

Virtue ethics shifts focus from rules or outcomes to the character of the decision-maker. It asks, **What kind of leader am I becoming through this decision?** Integrity, courage, humility, and fairness form the foundation.

For IT leaders, this means embodying values that shape not only choices but culture. Integrity resists shortcuts; humility invites multiple perspectives; courage defends the right path even when unpopular.

Integrate virtue-based evaluation into 360 reviews. Ask, **How did this leader show integrity, humility, and courage during moments of ambiguity?** Make ethical conduct observable and promotable.

Integrating Frameworks in IT Leadership

No single framework provides a complete answer, but together they create a balanced toolkit. Kant's deontology ensures leaders never compromise fundamental human rights. Utilitarianism allows flexibility to weigh broader societal outcomes. Virtue ethics centers the personal character and culture of leadership, ensuring consistency and integrity over time.

In practice, an IT leader might ask: **Does this action respect dignity and fairness? Does it maximize benefits and minimize harm? Does it reflect the kind of leader and organization we aspire to be?** Applying all three perspectives, decisions become not only ethically sound but also strategically resilient and deeply human-centered.

Establish an **Ethics Lens** in strategic planning documents or product requirement templates. Every major decision should explicitly show how it satisfies all **three pillars**: dignity, outcome equity, and personal integrity. This becomes part of how you work, not just how you react.

This integrated approach strengthens organizational credibility, reassures stakeholders, and fosters a culture where ethical reasoning is embedded into daily operations. Leaders who embrace multiple frameworks are better equipped to handle complexity, balance competing priorities, and inspire confidence across diverse audiences.

Ethical frameworks give IT leaders structured ways to approach difficult dilemmas where intuition alone may fall short. While there are many schools of thought, three stand out as especially relevant to technology and leadership: Kant's deontological ethics, utilitarianism, and virtue ethics. Each offers a distinct lens for decision-making, one rooted in principles and duties, another in outcomes and consequences, and the third in personal character and cultural influence. What follows is a summary of these three perspectives, highlighting their key ideas, applications in IT leadership, and their respective strengths and limitations.

Building an Ethical IT Culture

Ethical leadership begins with individuals but only through culture. A culture of ethics does not emerge by accident; it requires deliberate modeling and consistent reinforcement.

Integrate ethics visibly into daily operations, make it a standing agenda item in leadership meetings and project reviews. Create an **Ethics in Action Portal** where employees can log dilemmas or observe how leadership resolved real cases. Transform principles into living stories.

1. **Set the Tone at the Top:** Demonstrating honesty, transparency, and accountability establishes non-negotiable standards. Admitting mistakes publicly reinforces that ethical choices outweigh convenience. Host quarterly **Ethics Reflection Sessions** where executives discuss recent dilemmas and decisions, modeling visible candor.
2. **Integrate ethics into daily decisions:** Every strategy session or rollout should include: Does this align with our values? Who is impacted, and how do we safeguard them? Add a permanent line titled **Ethical Dependencies** in every project charter, identifying affected groups and protective measures.
3. **Empower employees to speak up:** They must feel safe raising concerns without fear. Establish open channels and rapid-response leadership reviews. Implement a digital **Speak-Up Dashboard** showing issue volume, response times, and resolutions to build trust.
4. **Reward Ethical Behavior:** Recognizing principled decisions signals that integrity equals performance. Introduce annual **Integrity Impact Awards** honoring employees who made difficult but value-driven calls that protected credibility.

Guiding Your Decisions with Ethical Principles

Ethical decision-making demands more than instinct; it requires a steady compass to navigate complex and ambiguous situations. Core principles such as transparency, accountability, and equity provide that compass, ensuring that leadership choices remain trustworthy and consistent even under pressure.

Operationalize these principles into repeatable steps. Use decision worksheets with clearly labeled ethical checkpoints. Make these reviews part of portfolio and budgetary governance, not side conversations.

Create a lightweight **Ethical Impact Canvas** one page per initiative capturing: affected stakeholders, data sensitivity, unintended-consequence analysis, and mitigation steps. Use it as a pre-launch requirement.

Build scenario-based training where teams role-play dilemmas involving conflicting stakeholder interests or data misuse risks. The more practiced your people are, the more intuitive ethical reasoning becomes in fast-moving contexts.

Run these simulations quarterly using anonymized real incidents from your sector; rotate facilitators so every manager gains skill in guiding ethical debate.

1. **Transparency:** strengthens trust through open communication, publishing FAQs, clarifying consent, and admitting vulnerabilities before rollout.
2. **Accountability:** reinforces integrity by owning outcomes and sharing lessons quickly.
3. **Equity:** ensures fairness in design and execution, demanding bias testing and inclusive review.
4. **Privacy and Security:** uphold rights by limiting data use to what is essential.
5. **Sustainability:** secures the future by embedding carbon and e-waste considerations into procurement and infrastructure decisions

Together, these principles turn ethics from abstract ideals into practical standards that guide decisions, strengthen credibility, and build cultures rooted in trust.

Navigating Ethical Dilemmas

In IT leadership, you will inevitably face ethical dilemmas, situations where the right course of action isn't immediately clear. To navigate these challenges, you must approach them with a structured mindset.

Adopt a five-question decision model, **what values are at stake, who is affected, what principles apply, what trade-offs exist, and what precedent is set**. Document answers before proceeding.

Integrate this analysis into workflow approvals so reflection becomes procedural, not optional.

Create safe escalation paths. Teams must feel empowered to raise red flags early. Leaders should welcome these discussions as signals of integrity.

Establish **Ethics Circles** monthly open forums under confidentiality where employees can present dilemmas for joint problem-solving.

Once dilemmas are resolved, [publish concise internal **Ethics Memos** summarizing the reasoning and lessons learned so reflection becomes normalized practice.

1. **Define the Problem Clearly:** Identify the core ethical issue and the stakeholders involved. Ask yourself what's truly at stake and who will be impacted by your decision.
2. **Evaluate the Options:** Consider all possible courses of action. Analyze their potential outcomes, weighing the benefits and risks for all stakeholders. Look beyond financial metrics to assess reputational, social, and environmental implications.
3. **Seek Diverse Perspectives**: Involve colleagues from different backgrounds, roles, and departments to provide fresh insights. Ethical decisions are rarely made in isolation, and diverse viewpoints can help you avoid blind spots.
4. **Consult Ethical Frameworks**: Use established frameworks like the **Kantian** approach (focusing on principles and duties) or utilitarianism (maximizing overall good) to guide your thinking.
5. **Decide with Integrity:** Once you've weighed the options, make a choice that aligns with your values and principles. Even if it's not the easiest or most profitable option, prioritize what is right over what is convenient.
6. **Communicate Your Decision:** Be transparent about your reasoning and the factors you considered. Share your thought process with stakeholders to demonstrate your commitment to ethical leadership.

   After each major decision, publish a short internal **Ethics Memo** summarizing how the decision met ethical checkpoints. Normalize visible reflection.

Harnessing Technology for Good

Ethical IT leadership isn't just about avoiding harm in actively using technology to create positive change. As a leader, you have the power to direct technological advancements toward solving meaningful problems and improving lives.

Make ethical innovation a core investment area. Prioritize projects that support accessibility, reduce carbon impact, or expand digital inclusion. Tie innovation goals to measurable ethical outcomes, not just features or cost savings. When evaluating partnerships or vendors, it goes beyond functionality. Assess their data practices, labor policies, and environmental standards. Build procurement scorecards that reflect your ethical stance, and act on them.

Establish a **Tech for Good**" portfolio within your roadmap, set aside 10 percent of annual innovation budget for projects that deliver measurable social or environmental benefit.

1. **Develop Responsible AI:** Ensure that your AI systems are free from bias, transparent in their decision making, and aligned with ethical values. Regularly audit your algorithms to prevent unintended consequences. Use third-party fairness-testing APIs and publish audit summaries to your governance board quarterly.
2. **Champion Accessibility:** Design technologies that are inclusive and accessible to all users, regardless of their abilities or circumstances. Remove barriers that might prevent people from benefiting from your solutions. Adopt WCAG 2.2 standards as baseline; run semi-annual accessibility sprints reviewing all public interfaces.
3. **Drive Sustainability:** Leverage IT to reduce environmental impact. From optimizing energy use in data centers to promoting remote work, finding ways to align innovation with sustainability goals. Create a **Green Ops** dashboard measuring carbon savings per digital initiative; include it in board-level reporting.
4. **Protect Vulnerable Populations**: Use technology to address societal challenges like inequality, education gaps, or healthcare access. Focus on initiatives that empower underserved communities. Partner with NGOs or civic-tech organizations to pilot inclusion solutions; measure reach and social ROI annually.

Harnessing technology for good requires vision, responsibility, and courage. When IT leaders commit to building solutions that are ethical, inclusive, and sustainable, technology becomes more than a tool, it becomes a catalyst for progress that uplifts people, strengthens communities, and shapes a more equitable future.

Measuring and Maintaining Ethical Progress

To ensure that your ethical leadership is effective, you must regularly measure and evaluate its impact. Without monitoring, even the best intentions can lose momentum. Include ethics KPIs in leadership scorecards. Track indicators like ethical training completion, incident escalation response time, fairness audits on algorithms, and sentiment from employee trust surveys. Conduct 'ethics retrospectives' after major projects. Document what went well, where ethical tensions arose, and what could be improved next time. Institutionalize learning, not just compliance. Publish an annual **Ethical Progress Report** internally, highlight metrics, gaps, and future commitments, mirroring sustainability reporting standards.

1. **Define Clear Metrics:** Identify key indicators of ethical performance, such as data privacy compliance rates, diversity metrics, or customer trust scores. Use these metrics to track progress over time.
2. **Conduct Ethical Audits:** Periodically review your organization's practices to ensure alignment with your ethical principles. Look for areas where improvement is needed and act decisively to address gaps.
3. **Engage Stakeholders:** Regularly gather feedback from employees, customers, and partners to understand how your decisions are perceived. Use their input to refine your approach.
4. **Stay Educated**: Ethics is an evolving field, especially in the fast-changing world of IT. Stay informed about emerging challenges, regulatory updates, and best practices to ensure your leadership remains relevant. Subscribe your leadership team to ethics-in-technology consortiums (e.g., IEEE Ethically Aligned Design, Partnership on AI) to stay ahead of evolving standards.

Measuring and maintaining ethical progress is not a one-time exercise but a continuous commitment. Through consistent evaluation, stakeholder engagement, and ongoing education, IT leaders ensure that ethical principles remain alive in practice, driving trust, accountability, and long-term organizational resilience.

The Ethical IT Leader's Legacy

Your ethical choices today will define the future of your organization and the lives it touches. Leading with integrity, transparency, and purpose, you not only protect your organization from risks but also inspire others to follow your example. Your legacy as an IT leader will be measured not just by the technologies you deploy but by the trust you build, the fairness you ensure, and the impact you create.

Ethical leadership is not a one-time effort; it's a lifelong commitment. Every decision, every interaction, and every innovation is an opportunity to demonstrate your values. Over time, these choices accumulate, shaping a culture where accountability and respect become non-negotiable standards. Future leaders will look to your example as proof that progress, and principle can coexist. Embedding ethics into the foundation of your organization, you leave behind more than successful projects, you leave a blueprint for responsible innovation.

Seize these opportunities with courage and clarity, knowing that your leadership is shaping a better, more equitable world. The path may not always be easy, but it is always worth it. **Finalize your ethical legacy plan**: write your principles, codify them into leadership onboarding, make ethics part of promotion criteria, and publish your personal ethics statement. This cements accountability into succession.

Start now. Build your ethics dashboard. Write your principles. Ingrain them in hiring, reviews, and operations. Lead every major initiative with ethics at the core. These are the steps that transform ethical leadership from ideal to reality, and from reality to legacy.

Chapter 12

The Global Visionary IT Leader

Global IT leadership demands more than technical expertise; it calls for a sophisticated balance of cultural intelligence, strategic vision, and the ability to drive collaboration across diverse geographies. As technology transcends borders, the role of an IT leader expands to become a unifying force that aligns local realities with global ambitions. To lead effectively on this scale, you must cultivate a mindset that thrives in complexity, embraces inclusivity, and fosters innovation.

Document your global leadership operating rhythm. Define how often you engage in each region, how input is collected, and how global goals are re-grounded in local insights. Make consistency the foundation of clarity.

Global leaders must also recognize the ethical and social responsibilities that come with shaping technology across different societies. Decisions made in one region can influence economic stability, security, and opportunity in another. Building trust across cultures requires humility, adaptability, and the courage to respect diverse perspectives. A true visionary IT leader is not only a strategist but also a bridge-builder, creating environments where diverse talent can flourish, and ensuring that technology serves as a force for shared progress worldwide.

Create a cross-regional **Voices and Impact** series. Feature brief narratives from different teams highlighting how local work is creating global change. Show your teams they are building more than systems, they're building progress.

This chapter provides a comprehensive guide to mastering the art of global IT leadership, equipping you with the tools and insights to build influence, inspire teams, and drive meaningful impact across the interconnected world.

8 PRINCIPLES OF THE GLOBAL VISIONARY IT LEADER

1 BALANCE EXPERTISE & VISION
Lead with both technical mastery and strategic foresight across geographies.

2 EMBRACE CULTURAL INTELLIGENCE
Respect regional diversity, adapt to norms, and build global trust.

3 UNIFY LOCAL & GLOBAL GOALS
Align regional realities with shared organizational ambitions.

4 NAVIGATE REGULATIONS WISELY
Balance innovation with compliance in privacy, security, and ethics.

5 EMPOWER DISTRIBUTED TEAMS
Foster inclusion, autonomy, and collaboration across borders.

6 DRIVE CROSS-BORDER INNOVATION
Leverage global creativity and data to deliver transformative solutions.

7 ADAPT WITH AGILITY
Continuously evaluate metrics, recalibrate, and evolve strategies.

8 LEAD WITH PURPOSE & PASSION
Inspire with vision, empathy, and a legacy of impact.

A TRUE GLOBAL IT LEADER UNITES VISION, ETHICS, AND INNOVATION TO SHAPE PROGRESS THAT TRANSCENDS BORDERS.

Redefining Global Leadership in IT

Stepping into the role of a global IT leader involves more than overseeing multinational operations. It is about weaving a unified vision that respects regional nuances while advancing cohesive organizational objectives. Leadership in this domain requires an openness to new ways of thinking and a commitment to learning from diverse cultures and approaches.

The foundation of global leadership lies in adaptability. Your leadership style must be versatile enough to resonate with individuals from varying cultural and professional backgrounds. For example, decision making norms differ widely, some cultures value consensus driven processes, while others lean toward swift, top-down authority. Recognizing and adjusting to these differences demonstrates respect and builds trust.

Additionally, global leadership means recognizing that technology serves as both a unifier and a differentiator. The tools you implement must foster collaboration across regions while accommodating local needs. Your ability to align these elements determines the success of your global vision.

Set a standard **Local Fit Diagnostic** for major platform rollouts. Before deployment, regional teams assess whether the initiative meets local needs in UX, compliance, and business model alignment. Adjust and localize where necessary before global scale.

To lead effectively, implement a repeatable framework for evaluating global initiatives through local lenses. When introducing enterprise-wide platforms or policies, task regional advisors with stress-testing the rollout for relevance and friction. Require leaders to document how both local and global goals are being supported.

Building Cultural Intelligence

Cultural intelligence is a cornerstone of effective global leadership. Without it, even the most well-intentioned initiatives can falter due to misunderstandings or misaligned priorities.

To develop cultural intelligence, immerse yourself deeply in the cultures you aim to lead. This involves more than a superficial understanding of traditions or etiquette; it requires a grasp of how people communicate, solve problems, and make decisions. For instance, in high context cultures, much of the communication may be implied rather than directly stated, requiring you to read between the lines and interpret subtle cues.

Create opportunities for your team to share their cultural insights openly. This dialogue not only enhances your understanding but also fosters a sense of inclusion and respect. Additionally, leverage tools like cross cultural training programs and multilingual platforms to address communication barriers and build stronger connections across teams.

Ultimately, cultural intelligence is about listening with intention, learning with humility, and leading with empathy. When your team sees your commitment to understanding their world, they are more likely to embrace your vision.

Incorporate cultural briefings as part of all onboarding and project ramp-ups. Equip leaders and teams with a regional snapshot, norms, risk sensitivities, work rhythms, to enable smoother engagement.

Embed cultural intelligence checkpoints into project lifecycle stages, especially during early design and user adoption planning. Assign regional cultural ambassadors to review narratives, onboarding materials, or change communications for relevance. Don't just translate, localize.

Crafting a Unified Vision

A global leader's vision must bridge the gap between diverse local realities and the broader organizational mission. It is not enough for your vision to be compelling; it must be actionable and relevant to all stakeholders, regardless of their location or context.

Begin by identifying universal values that resonate across the organization, such as customer centricity, innovation, or sustainability. These values form the backbone of your vision and create a shared purpose that transcends regional differences. For instance, while specific markets may prioritize cost efficiency or technological advancement, the overarching focus on delivering customer value remains constant.

Once your vision is articulated, communicate it with precision and passion. Use stories and real-world examples to illustrate its impact. Highlight successes from different regions to show how each team contributes to the larger mission. Tailor your messaging to suit the audience, acknowledging regional challenges and opportunities while emphasizing their alignment with the global strategy.

Operationalize your global vision through cascading OKRs (Objectives and Key Results) that connect top-level strategy to local execution. Ensure each regional team can trace their objectives directly to the broader vision. This creates alignment without mandating uniformity.

Use a **Vision Traceability Matrix** to link every regional program to a strategic global outcome. Update it quarterly, and publish excerpts to reinforce cohesion.

Navigating Global Regulations and Risks

The global stage presents IT leaders with the dual challenge of innovation and compliance. Operating across borders means engaging with an intricate web of regulations that govern data privacy, cybersecurity, and digital operations. These laws, while essential to protecting individuals and systems, vary greatly across regions, creating a complex landscape that demands vigilance, foresight, and adaptability. Navigating these regulations is not simply a matter of ticking legal checkboxes; it requires strategic integration of compliance into the fabric of IT leadership.

Build an internal policy clearinghouse where global and regional legal, risk, and IT leaders can collaborate on regulatory reviews before launching any new product or data-driven initiative. Document decisions clearly with business rationale and local impact.

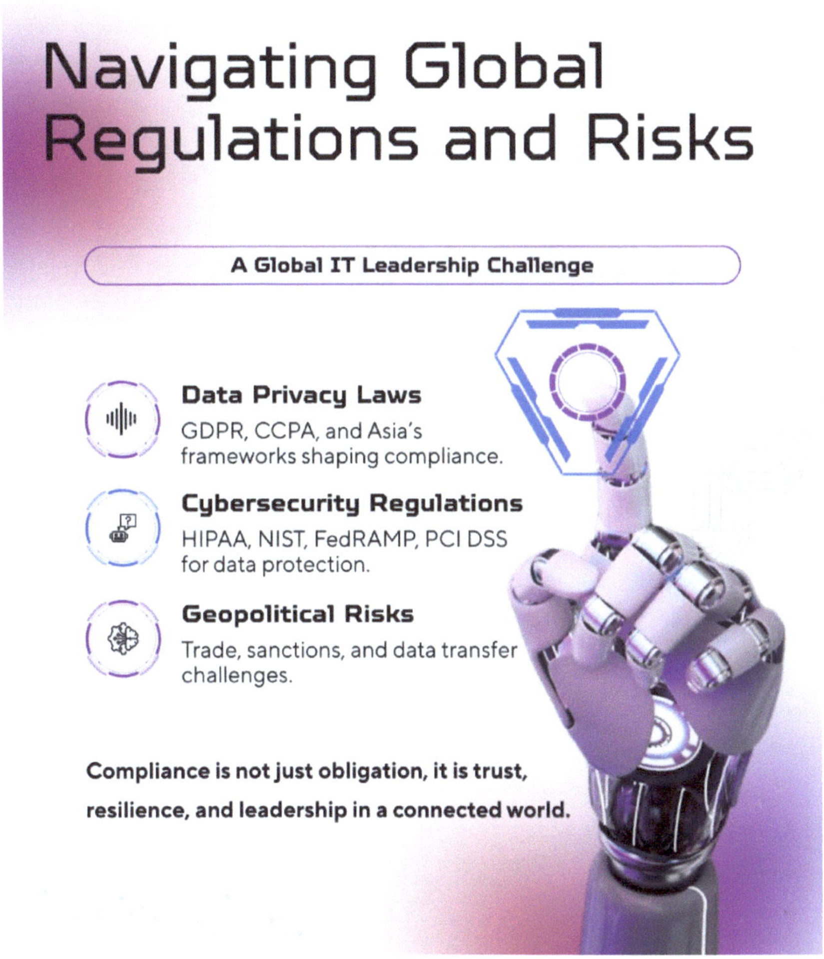

At the forefront of this regulatory maze are data privacy laws, which have become increasingly stringent in the digital age. The European Union's General Data Protection Regulation (GDPR) sets a global benchmark with its robust requirements for data protection, individual consent, and accountability. Similarly, the California Consumer Privacy Act (CCPA) emphasizes consumer rights and transparency, reflecting the growing demand for privacy in the United States. Asia adds another layer of complexity with China's

Cybersecurity Law and emerging frameworks in regions such as India and Japan. Each regulation mandates specific standards for data collection, storage, and sharing, requiring leaders to customize compliance strategies for each market.

In addition to privacy, cybersecurity regulations are critical. The United States enforces laws like the Health Insurance Portability and Accountability Act (HIPAA) for healthcare data and the Federal Risk and Authorization Management Program (FedRAMP) for cloud services. These frameworks prioritize securing sensitive information against breaches. Across industries, sector specific rules such as the Payment Card Industry Data Security Standard (PCI DSS) ensure that financial data is handled securely. For IT leaders, compliance is not just a legal mandate; it is a trust building exercise that reassures stakeholders of the organization's commitment to safeguarding their information.

The challenge does not stop implementing policies; it extends to keeping pace with evolving standards. Regulatory landscapes shift frequently in response to technological advancements, high profile breaches, and shifting public expectations. For example, recent discussions around AI ethics are likely to result in new global standards for algorithm transparency, bias mitigation, and accountability. IT leaders must actively monitor these developments to avoid being caught off guard by unexpected requirements.

The stakes are further elevated by geopolitical dynamics, where trade policies, economic sanctions, and cross border data transfer agreements can significantly impact operations. For instance, restrictions on data flows between regions such as the EU and the US necessitate the adoption of standard contractual clauses or other mechanisms to ensure lawful compliance. Additionally, political uncertainties can disrupt supply chains or lead to fluctuating regulations that require rapid adaptation.

Addressing these challenges requires an integrated approach. Building a governance framework that balances global consistency with local flexibility is essential. Such a framework should align the organization's overarching objectives with the specific demands of each market. Partnerships with legal and compliance experts are invaluable in interpreting nuanced regional laws and anticipating changes. Proactive risk assessments and audits should be conducted regularly to identify gaps and strengthen the organization's compliance posture.

Beyond adherence, effective communication about compliance efforts builds trust with customers, partners, and regulators. Transparency regarding data practices, security protocols, and ethical considerations positions your organization as a responsible global player. Whether it's publishing privacy policies, issuing regular compliance updates, or engaging in industry discussions on emerging regulations, being visible and proactive underscores your commitment to ethical operations.

In the ever-evolving regulatory landscape, staying ahead is not merely about avoiding penalties but about positioning your organization as a leader in responsible technology use. This requires vigilance, collaboration, and the ability to embed compliance into the organization's strategy seamlessly. As an IT leader, your ability to navigate these challenges while driving innovation is a defining element of your global success.

Assign a rotating compliance watch team drawn from IT, legal, data, and product to track developments and advise on high-priority regulatory risks quarterly. Publish short internal updates that translate legal shifts into operational impact.

Create **Compliance Radar** newsletter, brief, action-oriented updates highlighting changes in data law, cross-border constraints, and AI regulation shifts. Pair each item with a playbook action.

Empowering Distributed Teams

The strength of your leadership lies in your ability to empower teams spread across the globe. Distributed teams bring unique challenges, such as time zone disparities, resource inequities, and cultural differences. However, with the right strategies, these teams can become a source of unparalleled innovation and agility.

Start by cultivating trust and inclusiveness. Make every team member feel valued by recognizing their contributions and providing equal opportunities for growth. Celebrate milestones on both regional and global levels to reinforce a sense of shared purpose.

Invest in technologies that bridge physical divides. Real time collaboration tools, project management platforms, and virtual meeting software create spaces for seamless interaction. Beyond technology, prioritize personal connections. Regular visits to regional offices foster stronger relationships and provide firsthand insights into local challenges and achievements.

Encourage autonomy by empowering regional leaders to make decisions relevant to their markets. This not only accelerates responsiveness but also fosters a sense of ownership that drives commitment and innovation.

Establish a global team enablement charter that outlines how teams will collaborate, escalate issues, share learnings, and influence decisions regardless of geography. Adopt rotating time zone leadership for critical meetings to show respect for distributed teams.

Track inclusion in decision-making through simple pulse surveys.

Introduce a **Global Voice Index** in your engagement surveys, track perception of influence, decision proximity, and visibility for distributed teams.

Driving Innovation Across Borders

Innovation is the lifeblood of global IT leadership. Your role is to harness the creativity and expertise of your teams across regions to develop transformative solutions.

Create innovation hubs or centers of excellence in strategically chosen locations, where teams can experiment with emerging technologies and design solutions tailored to their markets. Facilitate cross regional collaboration by organizing global hackathons, innovation challenges, and knowledge sharing sessions that bring diverse perspectives together.

Data is a powerful tool for fostering innovation. Use analytics to identify trends, anticipate customer needs, and uncover opportunities for growth. Whether it's improving supply chain efficiency or developing personalized customer experiences, data driven decision making enables you to stay ahead in a competitive global landscape.

Institute a global knowledge-to-value loop: encourage each region to document insights gained from pilots, share them via structured reviews, and convert them into global learning assets. Use a centralized innovation intake board to surface ideas and co-fund promising prototypes with multi-region benefit.

Run quarterly **Innovation Roundups** across all hubs, short, focused sessions where teams pitch lessons learned and failed experiments alongside wins. Normalize transparent experimentation.

Measuring Impact and Adapting for the Future

Global leadership requires continuous evaluation and adaptability. Establish metrics that reflect both regional and global performance, such as operational efficiency, customer satisfaction, and employee engagement. Regularly review these metrics to assess progress and identify areas for improvement.

Be open to recalibrating your strategies based on new insights or shifting circumstances. For instance, if market conditions change in a key region, be prepared to adjust your approach to maintain alignment with local realities while staying true to the overarching vision.

Augment global dashboards with regional-specific KPIs that ladder up into enterprise outcomes. Make it a quarterly practice to review which metrics remain relevant, and sunset those that no longer drive insight. Require each global function to document one adaptation they made based on metrics every cycle.

Embed a **Metrics-to-Actions** discipline, track not just data, but specific leadership decisions triggered by those metrics. Use as input in regional strategy reviews.

Leading with Purpose and Passion

At its heart, global IT leadership is about people. It's about fostering connections, inspiring collaboration, and creating value that transcends borders. The impact of your leadership extends far beyond technology; it influences the lives of employees, customers, and communities worldwide.

Lead with a clear sense of purpose, allowing that purpose to guide every decision you make. Stay passionate about your role, and use that passion to inspire those around you. Doing so, you will not only achieve success for your organization but also leave a lasting legacy of innovation, inclusivity, and impact.

Remind yourself and your teams regularly why the work matters, anchor project kickoffs in purpose statements, revisit global commitments in internal briefings, and use stories from the field to reinforce the human impact behind technical work. Passion and purpose scale when they're made visible and repeatable.

Create a library of **impact narratives,** brief, localized stories showing how global tech decisions improved lives. Use them to open all-hands meetings and reinforce why the mission matters.

The journey of a global visionary IT leader is filled with challenges and opportunities. Embrace them with confidence, empathy, and strategic intent. The stage is your step into it with purpose and conviction.

Leading with Purpose and Passion

Global IT leadership is about people, fostering connections, inspiring collaboration, and creating value across borders. Lead with clarity and passion, and your legacy will be innovation, inclusivity, and lasting impact.

Chapter 13

Driving Sustainability Through IT

The environmental challenges facing our planet demand leadership that aligns technological advancement with ecological responsibility. IT leaders are uniquely positioned to leverage technology to drive meaningful sustainability initiatives. This role extends beyond compliance or public relations is about embedding ecological stewardship into every facet of technology strategy. Sustainability in IT is a commitment to creating a future where innovation and environmental responsibility coexist harmoniously, delivering measurable impact for both the organization and society.

Adopt this as a mindset: position sustainability as a core strategic pillar, not a separate track or a compliance effort. Bake it directly into platform roadmaps, architecture decisions, vendor evaluations, and workforce upskilling programs.

Understanding the Relationship Between IT and Sustainability

Sustainability in IT requires recognizing the dual role of technology as both a contributor to environmental strain and a powerful tool for change. The energy demands of data centers, the growing issue of electronic waste, and the environmental cost of rapid technological development have made sustainability an essential consideration for IT leaders. Addressing these challenges is not just a moral obligation but a strategic decision that benefits reputation, operational efficiency, and stakeholder engagement.

An IT organization that integrates sustainability into its practices can achieve cost savings, meet regulatory expectations, and appeal to environmentally conscious customers and investors. Beyond these tangible benefits, this alignment creates a lasting cultural impact, demonstrating the organization's dedication to innovation with a purpose.

Create an annual sustainability tech audit to assess your systems across energy usage, e-waste, cloud efficiency, and data retention. Track cost, carbon footprint, and policy alignment in the same dashboard. Use these audits to anchor your sustainability strategy in tangible data.

Driving Sustainability Through IT

Align innovation with ecological responsibility to create lasting impact for organizations and society.

Transforming Data Centers into Green Operations

Data centers remain some of the most energy intensive components of IT infrastructure. Their optimization is central to driving sustainability. Leaders must take proactive steps to make these facilities more energy efficient and environmentally friendly.

Begin with a detailed energy audit to identify areas of inefficiency. Advanced cooling systems, such as liquid cooling and artificial intelligence driven airflow management, can significantly reduce energy consumption by tailoring cooling to real time needs. These systems not only lower costs but also extend the lifespan of hardware by preventing overheating.

A shift toward renewable energy can amplify these efforts. Collaborate with providers of solar, wind, or hydroelectric energy to sustain power data centers sustainably. Many leading organizations, including major cloud providers, have successfully transitioned to renewable energy sources. Achieving this milestone signals to stakeholders that your organization prioritizes long term sustainability.

Virtualization also holds tremendous potential for reducing energy use. Consolidating servers and minimizing underutilized resources, virtualization optimizes energy efficiency while reducing the overall hardware footprint. Ongoing monitoring of these systems ensures they operate at peak efficiency and provides valuable data to guide future improvements.

Use an energy-ROI playbook when making data center investments, compare not just operational cost, but energy recovery rates, long-term carbon offset value, and sustainability impact per workload migrated or virtualized.

Track PUE (Power Usage Effectiveness) quarterly, and tie executive compensation or team incentives to hitting green benchmarks. Report this at board level to elevate visibility.

Addressing the Electronic Waste Crisis

The rapid pace of technological innovation has exacerbated the global issue of electronic waste. IT leaders play a critical role in mitigating this problem by extending the lifecycle of devices, implementing responsible recycling practices, and promoting circular IT strategies.

Extending hardware lifespan begins with proper maintenance and regular updates. Upgrades, such as replacing traditional drives with solid state alternatives or adding memory, can revitalize devices and delay their obsolescence. These measures reduce the volume of discarded equipment and maximize the return on investment for hardware.

At the end of a device's lifecycle, engage with certified recycling partners who adhere to stringent environmental and ethical standards. Recycling not only reduces waste but also recovers valuable materials, such as precious metals, that can be reused. Donation programs are another effective strategy. Repurposing devices for educational or community use demonstrates a commitment to social responsibility while minimizing environmental impact.

Formalize a **Device Circularity Policy**. Require all asset refresh plans to evaluate: **Can this device be upgraded? Can it be donated? Can it be harvested for reuse?** Build procurement scorecards that reward vendors who support refurbishing and take-back programs.

Embracing Cloud Solutions for Environmental Efficiency

Cloud computing has emerged as one of the most impactful tools for sustainability in IT. Migrating operations to the cloud allows organizations to benefit from shared infrastructure, which significantly reduces energy consumption compared to on premises systems.

Selecting a cloud provider with strong environmental credentials is essential. Many providers now publish detailed sustainability metrics, offering transparency about their energy use and carbon reduction strategies. These insights allow organizations to make informed decisions and align their technological goals with environmental priorities.

Once cloud systems are implemented, resource optimization becomes the next focus. Dynamic scaling ensures that computing resources are allocated efficiently, avoiding waste. Load balancing and automated workflows further enhance energy efficiency, demonstrating that sustainability and performance can go hand in hand.

During cloud vendor reviews, add **Green Score** evaluation criteria, based on third-party carbon certifications, datacenter renewable use, and published emissions targets. Push for shared responsibility in sustainability SLAs.

Automate idle instance monitoring, enforce time-limited development environments, and enable cost-governance tools as part of FinOps practices to align cloud cost savings with emissions reduction.

Leveraging IoT for Sustainable Operations

The Internet of Things presents unparalleled opportunities to improve resource efficiency across industries. Integrating IoT solutions, IT leaders can create smarter, more sustainable operations.

Energy management systems equipped with IoT sensors monitor power consumption in real time, identifying areas for improvement. For example, automated lighting and heating systems can adjust based on occupancy, reducing unnecessary energy use. These small adjustments add up to significant savings over time while demonstrating environmental responsibility.

Logistics is another area where IoT can make a difference. Sensors embedded in fleet vehicles optimize delivery routes, monitor fuel efficiency, and minimize emissions. These data driven approaches reduce costs and contribute to sustainability goals, providing quantifiable results that reinforce your organization's commitment to responsible practices.

Establish a Green IoT Governance Framework. Require environmental justification for each sensor deployment. Monitor environmental return (carbon or cost saved per unit) and remove outdated or inefficient devices.

Creating a Culture of Sustainability

Sustainability initiatives are most effective when they are embraced at every level of the organization. Cultivating a culture that values and prioritizes ecological responsibility begins with education.

Equip your teams with the knowledge and tools to make environmentally conscious decisions. Workshops and training sessions provide practical guidance, while open forums encourage the sharing of innovative ideas. Empower employees to suggest and implement changes, fostering a sense of ownership in the organization's sustainability journey.

Transparency is vital for maintaining engagement. Share regular updates about progress toward sustainability goals, highlighting both successes and areas for improvement. This not only builds trust but also motivates teams to continue contributing to the organization's mission.

Include sustainability metrics in all major project retrospectives. Celebrate teams who reduce waste, retire legacy tech, or rearchitect to optimize efficiency. Normalize it as a mark of engineering excellence.

Measuring and Communicating Success

Sustainability efforts must be backed by rigorous measures to ensure progress and identify opportunities for growth. Key metrics, such as energy consumption, carbon emissions, and waste reduction, provide valuable insights into the effectiveness of your initiatives.

Create accessible dashboards that offer a real time view of these metrics. Sharing this data with stakeholders, from employees to investors, reinforces your commitment to accountability and continuous improvement. Publicly communicating achievements through reports, social media, or press releases amplifies the impact of your efforts, inspiring others to follow suit.

Define a sustainability success index composed of carbon reduction, energy savings, device reuse, and cloud efficiency. Tie these KPIs to the IT operating model, not just CSR reporting. Use them in quarterly business reviews with leadership.

A Leadership Imperative

Sustainability is not a peripheral concern for IT leaders, it is a core component of effective leadership in the modern era. Driving sustainability through IT, you are not only addressing immediate environmental challenges but also positioning your organization for long-term success.

The actions you take today lay the groundwork for a future where innovation and responsibility are inseparable. Whether through optimizing data centers, reducing electronic waste, or fostering a culture of environmental stewardship, your leadership has the power to create lasting impact. Embrace this opportunity to lead with vision and purpose, ensuring that technology serves as a force for good in shaping a sustainable world.

Use every large transformation effort, cloud migration, ERP modernization, hardware refresh as a trigger to layer in sustainability commitments. Ask, every time: **How can we reduce, reuse, or rearchitect this in greener ways?** Make that a default expectation.

Part 5

Engaging and Inspiring Others

(Chapters 14 - 15 - 16)

Part 5 focuses on the critical role IT leaders play in inspiring and connecting with teams, stakeholders, and customers through meaningful engagement and leadership practices.

In **Chapter 14**, IT's Role in Customer Experience Excellence, leaders explore how technology can create seamless, personalized, and memorable customer journeys. This chapter emphasizes the importance of integrating systems, leveraging data, and aligning IT with customer centric strategies.

Chapter 15, Storytelling as a Leadership Tool in IT, highlights the power of compelling narratives to inspire action, build trust, and forge connections. Leaders learn how to use storytelling to translate complex technical concepts into relatable ideas that resonate with diverse audiences.

In **Chapter 16**, Mentorship in IT Leadership, the focus shifts to nurturing future talent. This chapter provides actionable strategies for guiding the next generation of IT leaders, emphasizing the transformative impact of mentorship on both individuals and organizations.

Through these chapters, Part 5 empowers IT leaders to engage deeply, inspire confidently, and foster enduring connections that drive success.

Chapter 14

IT's Role in Customer Experience Excellence

Customer experience, or CX, has emerged as the defining factor in how businesses engage, retain, and expand their customer base. The expectations of seamless, intuitive, and personalized experiences are no longer aspirational; they are foundational to success. The task of meeting these expectations increasingly rests on the shoulders of IT leadership. Technology moves beyond a supporting role in CX, serving as the engine that drives it forward. IT leaders must transition from being back-office problem solvers to visionaries who actively shape the customer journey, ensuring every interaction aligns with customer desires and organizational goals.

Approach CX as a value-driving product in itself, one that needs ownership, outcome tracking, and evolution roadmaps, just like any mission-critical platform. Embed this thinking into your operating model.

IT's Role in Customer Experience Excellence

Empowering seamless, personalized, and connected customer journeys through technology-driven innovation.

Why Customer Experience is the New Currency

The experiences customers have with a brand carry as much weight, if not more, than the products or services they purchase. A product's quality or competitive pricing can draw customers in, but it is the experience that keeps them coming back or drives them to competitors. A frustrating or disjointed interaction can undo years of trust, while a delightful experience can create loyal advocates.

IT leadership sits at the heart of these experiences. Every digital touchpoint, from a mobile app to a chatbot or an in-store kiosk, relies on technology to function seamlessly. Beyond functionality, these touchpoints must be emotionally resonant, addressing customer needs intuitively. IT leaders are tasked with crafting systems that not only function well but also anticipate and exceed customer expectations. This involves harnessing technology to create solutions that are as engaging as they are efficient. Set up quarterly CX audits across your digital stack. Evaluate each interaction point not just for uptime or speed, but for human experience, friction points, and emotional clarity. Rotate team leads to shadowing real customers or support agents weekly to expose blind spots.

The IT Leader's Mindset for CX Excellence

Shaping exceptional customer experiences begins with a mindset shift. IT leaders must see themselves as more than technology managers; they are designers of the customer journey. This means adopting a customer first approach, asking questions like, "**What does the customer feel at this stage of the interaction?**" and "**How can technology make this moment more seamless, more personal, or more memorable**?"

This mindset requires stepping into the customer's world. Understanding their behaviors, frustrations, and aspirations, IT leaders can identify gaps in current processes and opportunities for innovation. It also involves fostering collaboration with marketing, sales, and customer support teams to ensure that technological solutions are deeply aligned with customer needs.

Install a **Voice of the Customer** loop inside product planning and release cycles. Require that at least one customer friction or feedback signal is presented alongside every backlog prioritization session.

Co-design with customers, not just for them. Pilot key digital journeys with real users during test phases. Codify this process into how you build.

Building Seamless Omnichannel Experiences

Today's customers expect their interactions with a brand to feel unified, whether they occur online, on mobile, or in a physical location. Achieving this level of consistency requires integrating technology systems across all touchpoints.

One of the primary challenges lies in breaking down silos between different platforms and departments. Customer data is often scattered across various systems, making it difficult to provide a unified experience. IT leaders must prioritize the creation of centralized customer profiles that consolidate data from CRM platforms, website interactions, social media activity, and in store visits. These profiles provide a complete view of the customer, enabling teams to deliver relevant and consistent experiences at every touchpoint.

For instance, if a customer begins browsing on a mobile app, adds items to their cart, and later visits a physical store, the store staff should have access to this context to assist in completing the transaction. This level of integration requires robust backend systems that ensure real time communication between digital and physical channels.

Create a CX Interoperability Charter. Mandate that each platform be interoperable with customer data layers. Run integration maturity assessments to close real-time sync gaps.

Use journey blueprints across key experiences (e.g., purchase, renew, return, support) and map technical systems to each step. Identify where disjointed handoffs occur, and define integration fixes as platform work, not customer support overhead.

Harnessing Personalization at Scale

Personalization has evolved from a nice to have feature to a critical component of CX excellence. Customers want to feel understood and valued, and IT leaders play a pivotal role in enabling these tailored interactions.

Personalization begins with data lots of it. Analyzing historical interactions, purchase behaviors, and preferences, IT teams can predict what customers are likely to want or need. For example, a streaming service might recommend new shows based on a user's viewing history, while an e commerce platform could suggest complementary products at checkout.

The challenge lies in delivering this personalization on a scale. AI and machine learning offer powerful tools to analyze data and generate personalized recommendations in real time. These technologies can dynamically adapt digital platforms to suit individual preferences, ensuring that every user feels like the experience was designed specifically for them.

However, with great data comes great responsibility. Customers are increasingly aware of how their data is used and demand transparency. IT leaders must implement strong data governance practices to ensure that personalization efforts respect privacy and comply with regulations.

Adopt a **Personalization Readiness Map**. Identify where your current tech stack supports contextual interactions, where it doesn't, and where you risk overreach or data misuse. Add progressive personalization capabilities into roadmaps, start with high-value, low-risk journeys.

Establish data ethics training for your engineering and analytics teams to reinforce the human impact behind every algorithmic decision.

The Power of Real Time Engagement

Modern customers expect immediacy. Whether resolving an issue or receiving updates, delays can lead to frustration and dissatisfaction. IT leaders must develop systems that enable real time engagement across all channels.

Chatbots powered by AI are a prime example of real time engagement in action. These tools provide instant answers to common queries, reducing waiting times and enhancing customer satisfaction. However, their effectiveness depends on integration with backend systems and access to up-to-date knowledge bases. IT leaders must ensure these bots are not only functional but also capable of contextual responses that feel human and empathetic.

Real time engagement extends beyond reactive support. Proactive notifications such as delivery updates, personalized promotions, or reminders demonstrate attentiveness and reduce uncertainty. For instance, an airline could notify passengers about gate changes or weather-related delays through automated systems, keeping them informed and reassured.

Define your Real-Time Capability Stack. Inventory what systems truly operate in real-time vs batch vs next day. Prioritize transformation projects that move your high-friction journeys (support, updates, transactions) into real-time pipelines.

Designing Digital Experiences That Captivate

The usability of digital platforms is often the first impression customers have of a brand. Websites, apps, and other interfaces must be visually appealing, intuitive, and efficient. Even minor frustrations, such as slow load times or confusing navigation, can deter users.

Regular performance audits are essential to maintain high quality digital experiences. These reviews should identify and address areas such as load speed, accessibility compliance, and overall design coherence. Mobile optimization is particularly critical, as a growing majority of customers interact with brands through their smartphones.

To stand out, incorporate emerging technologies like voice search and AI driven suggestions. These features simplify interactions and create a sense of modernity that resonates with tech savvy users. For example, allowing users to speak commands, such as "**Find nearby Italian restaurants**" enhances convenience and engagement.

Institute a Digital Experience Review Council, design, engineering, and CX meet monthly to score every digital product across experience quality dimensions. Require that each major release includes an experience benchmark uplift.

Leveraging IoT to Enhance Customer Interaction

The Internet of Things (IoT) opens new dimensions for customer interaction, offering opportunities to create connected, intelligent environments. From retail to hospitality, IoT transforms passive experiences into active, engaging ones.

Retail stores equipped with smart shelves can notify staff when stock levels are low or provide customers with instant access to product reviews through their smartphones. Similarly, hotels using IoT enabled room controls allow guests to personalize their environment by adjusting lighting, temperature, and entertainment systems with a single device.

However, deploying IoT solutions requires careful planning. IT leaders must ensure that these devices are integrated securely into existing networks and that the data they generate is used meaningfully. The focus should always remain on enhancing the customer experience, rather than overwhelming users with unnecessary complexity.

Create an IoT Interaction Blueprint. Map devices to touchpoints where they enhance, not clutter, the journey. Require CX use-cases to accompany all IoT deployment proposals.

Measuring the Impact of CX Initiatives

To improve CX, IT leaders must rely on both quantitative and qualitative insights. Metrics like Net Promoter Score (NPS), Customer Satisfaction (CSAT), and bounce rates reveal how customers perceive their experiences. However, these numbers only tell part of the story.

It's equally important to gather qualitative feedback through surveys, focus groups, and usability testing. These methods uncover the underlying reasons behind the metrics, providing actionable insights to refine systems and processes.

Regularly reviewing these metrics allows IT leaders to identify trends, address issues, and celebrate successes. Transparent reporting ensures that everyone in the organization remains aligned with CX goals. Use a **Customer Signal Repository** to store and tag insights across voice, chat, social, email, and field teams. Feed these insights into monthly product reviews and roadmap sessions. Require a response path for top recurring issues.

Cultivating a Customer Centric Culture

Exceptional CX is not the result of a single initiative but a cultural commitment to prioritizing the customer in every decision. IT leaders must champion this culture, demonstrating how technology solutions directly impact customer satisfaction.

Fostering collaboration between IT and other departments is essential. Sharing insights and aligning efforts across marketing, sales, and support teams creates a unified approach to CX. Recognizing and celebrating contributions to customer success reinforces the importance of this shared mission.

Implement a customer-impact reflection step in all major project reviews: **How did this release improve customer value?** Make this part of team performance rituals.

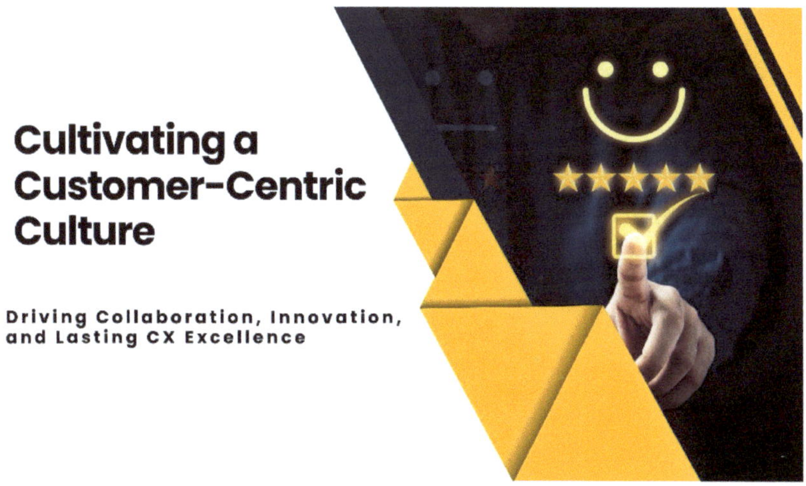

Leading the Charge in CX Innovation

The journey toward CX excellence is ongoing. It requires continual investment in learning, experimentation, and adaptation. IT leaders must stay informed about emerging technologies like augmented reality, virtual reality, and AI, exploring how they can be applied to enhance customer experiences.

Taking bold steps to innovate and consistently prioritize the customer, IT leaders can transform their organizations into benchmarks of CX excellence. The rewards include not only increased loyalty and advocacy but also a legacy of leadership that inspires others to follow.

The road to CX excellence is paved with opportunities. With vision, commitment, and a relentless focus on the customer, you can redefine what's possible and lead your organization to unparalleled success. The time to act is now, and the possibilities are limitless.

Run an annual CX Innovation Lab. Invite cross-functional teams to pitch disruptive improvements to top journeys. Fund top prototypes and fast-track them into active delivery. Let your organization practice CX reinvention regularly, not just during crises.

Chapter 15

Future Proofing IT Leadership

The future is unfolding faster than ever, shaped by technological breakthroughs, evolving market dynamics, and unforeseen global challenges. For IT leaders, **the question is not how to keep pace, but how to take the lead**. Future proofing IT leadership means building the capacity to anticipate, adapt, and excel amid constant change. It is about creating an enduring vision and equipping your organization to thrive in an uncertain landscape.

This chapter explores the strategies, skills, and mindset required to future proof IT leadership. It will guide you in identifying emerging opportunities, fostering resilience, and inspiring your team to rise to the challenges of tomorrow with confidence and creativity.

Leading with Vision and Foresight

Future proof leadership begins with clarity of vision and the ability to think strategically about what lies ahead. The most successful IT leaders are those who see beyond immediate challenges and envision how their decisions today will shape the future. This is not guesswork but a disciplined approach to studying trends, analyzing data, and making informed predictions.

Developing foresight requires continuous learning. You must immerse yourself in the forces shaping industries, from advancements in artificial intelligence to shifts in global economies. Stay informed by engaging with thought leaders, attending conferences, and collaborating with peers across industries. Cultivating a broad perspective, you can better understand how trends intersect and influence your organization.

Foresight is not a solitary endeavor. Encourage your team to think expansively, sharing their insights on emerging technologies and market developments. When you build a culture that values curiosity and exploration, your organization becomes a hub of innovation and preparedness.

As part of your leadership routine, implement what I call a "**Trend Horizon Review.**" Every 90 days, task your key team leads with surfacing one shift, technological, demographic, regulatory, or industry-specific, that could reshape operations in 18 to 36 months. Don't just review these findings, use them to trigger updates to your scenario planning, capability maps, and strategy roadmaps.

I also recommend creating a "**future signal repository**." This is a digital environment where your regional, departmental, or partner teams can contribute short entries capturing shifts they're observing, customer behaviors, new competitors, local disruptions. These signals, if shared early, become your organization's early warning system, not a postmortem collection.

Embracing the Role of Innovator and Architect

Innovation is the lifeblood of future proof leadership. As an IT leader, your role is to drive meaningful changes by challenging the status quo and reimagining what is possible. True innovation goes beyond incremental improvements; it transforms processes, products, and services to meet the needs of a changing world.

To innovate effectively, you must take risks thoughtfully. Identify inefficiencies and untapped opportunities within your organization, then design solutions that address these gaps. For instance, instead of simply automating existing workflows, consider how emerging technologies like machine learning or blockchain could redefine how work is done.

Being an architect of the future means building systems that are adaptable and scalable. Modular designs and interoperable technologies allow your organization to evolve without costly overhauls. Invest in IT infrastructures that can grow with your business, supporting new innovations and changing demands seamlessly.

Every innovation your team pursues should begin with what I call a **"Strategic Innovation Canvas."** This canvas outlines the long-range problem your innovation is solving, the legacy systems it may replace, and how success will look three years from now, not just after its first release. Using this framing forces your team to think beyond the short term.

Before any concept progresses to a formal initiative, it should be stress-tested through what I refer to as a **pilot-to-scale readiness review.** This review evaluates four readiness pillars: integration feasibility, measurable value potential, operational scalability, and risk mitigation (especially in security or compliance). If an idea fails in two or more pillars, it's not ready. Push for discipline before deployment.

Cultivating Organizational Agility

Agility is the defining characteristic of organizations that excel in uncertain environments. To cultivate agility, IT leaders play a key role in instilling a mindset and structure that allow teams to respond quickly and effectively to change.

Begin by fostering collaborative and cross functional teams. Break down silos and encourage open communication between departments, creating a unified approach to problem solving. Use frameworks like Agile or DevOps to enable iterative development, where teams can adjust their strategies in real time based on feedback and changing priorities.

Cultural agility is just as important as operational agility. Create an environment where experimentation is celebrated and setbacks are viewed as opportunities to learn and improve. When employees feel empowered to adapt and innovate, the organization becomes more resilient in the face of challenges.

I advise setting up what I call "**Fast Pivot Labs**" pop-up squads within your IT or cross-functional units designed to respond to fast-moving changes, competitive threats, or customer shifts. These labs should operate in two-week sprints and report impact weekly. The point is not perfection, it's speed-to-insight.

Don't limit Agile to software. Apply it to policy change, vendor strategy, and even employee development. Coach your business partners on how to run lightweight retrospectives and define their own backlogs for continuous improvement. This spreads adaptability across the enterprise.

Harnessing Emerging Technologies

Emerging technologies offer powerful tools for future proofing leadership, but their value lies in how effectively they are implemented. IT leaders take the initiative to stay ahead of technological advancements, not to chase trends, but to identify tools that align with their strategic goals.

Start by assessing which technologies hold the greatest potential for your industry. Artificial intelligence, for example, is transforming customer engagement through chatbots and predictive analytics, while advancements in cybersecurity are reshaping how organizations protect their data. Piloting these technologies on a small scale, you can evaluate their effectiveness and refine your approach before full scale implementation.

Adopting technology without a clear purpose risk wasting resources and diluting focus. Every technological investment should serve a defined objective, whether improving efficiency, enhancing customer experiences, or unlocking new revenue streams. A disciplined approach ensures that innovation aligns with organizational goals and drives measurable value.

What I've found most effective is introducing an **Emerging Tech Impact Grid** a simple heatmap updated every quarter that scores each new technology on four factors: likely time to adoption, relevance to your core strategy, regulatory exposure, and implementation complexity. This helps de-risk decisions and keeps conversations grounded in reality.

Once you've selected a few candidates from this grid, set up **use case sprints** short bursts where these technologies are tested in your real business context. Don't measure only outcomes, measure learning. Capture what worked, what failed, and what skills or data you lacked to fully extract value.

Building a Workforce for the Future

A future ready organization requires a skilled and adaptable workforce. As an IT leader, your responsibility extends beyond managing current talent to preparing your team for the challenges and opportunities of the future.

Begin by fostering a culture of continuous learning. Encourage employees to acquire new skills and explore emerging fields, whether through certifications, workshops, or collaborative projects. Provide access to training programs that align with organizational priorities, ensuring that your team remains at the forefront of technological expertise.

Automation and artificial intelligence are changing the nature of work, with some roles evolving and others becoming obsolete. Address these shifts proactively by identifying future skill requirements and offering reskilling opportunities. Support employees in transitioning to roles that emphasize strategic thinking, creativity, and problem-solving capabilities that machines cannot replicate.

Diversity within your workforce is another critical component of future proof. A team that brings diverse perspectives and experiences is better equipped to solve complex problems and innovate effectively. Strive to build an inclusive workplace where every voice is heard and valued.

Implement what I call a "**Skills Radar**." At the start of every quarter, bring together HR and business leaders to answer three questions: Which roles are evolving? **Which are in decline? Which are emerging faster than we expected?** Use this radar to shape internal mobility, talent acquisition, and learning investments.

Encourage job rotations and short-term project secondments as development tools. Let your data analyst spend 90 days inside customer experience. Let your infrastructure lead co-design an AI pilot. Don't just train people, let them experience the future by stepping into it.

Measuring Success and Adapting Strategies

Future proof leadership is a continuous process that demands regular assessment and refinement. Develop key performance indicators that reflect your long-term goals, from technology adoption rates to employee engagement and customer satisfaction.

Leverage analytics tools to monitor these metrics, identifying trends and areas for improvement. Use this data to make strategic decisions, ensuring that your efforts remain aligned with the organization's evolving needs.

Adaptability is central to future proofing. As market conditions shift or new challenges arise, revisit your strategies and adjust them as needed. Flexibility does not mean abandoning your vision; it means finding new paths to achieve it in the face of change.

Establish what I call a "**Future Fit Dashboard**" a blend of performance metrics and behavioral signals. Track things like innovation velocity, technology adoption cycles, cost-to-value ratios, learning progress, and adaptability scores. But also listen for softer signs: employee sentiment about change, team agility under stress, and how fast good ideas travel inside your culture.

Ask every functional leader to identify at least one policy, process, or system each quarter that no longer serves the future and submit a proposal to revise or retire it. This creates a rhythm of intentional renewal.

Inspiring Through Vision and Action

The most effective IT leaders inspire their teams by articulating a compelling vision and demonstrating the commitment to bring it to life. A future proof leader does not just react to change but leads with purpose, showing their organization what is possible.

Share your vision in a way that resonates with your team and stakeholders. Highlight how your strategies will create value, solve problems, and open new opportunities. Use storytelling to connect the dots between individual contributions and the broader mission, fostering a sense of ownership and engagement.

Actions speak louder than words. Model the behaviors you want to see in your organization by embracing innovation, practicing adaptability, and celebrating progress. When your leadership is rooted in both vision and execution, you build trust, loyalty, and momentum.

Tie your strategy updates to stories, not just data. Share real examples where someone in your organization spotted a weak signal, acted on it, and made a difference. Stories cut through noise, they make vision real.

When recognizing employees, don't just reward delivery, recognize foresight, adaptability, and learning. These are the muscles of future proof organizations.

Creating a Lasting Impact

Future proof leadership is not about short-term gains but about building a legacy of resilience, adaptability, and innovation. Leading with foresight, empowering your team, and embracing change, you position your organization to thrive in an ever-evolving world.

Your journey as an IT leader is one of constant growth and discovery. Each decision you make and every strategy you implement contributes to a future that is not only sustainable but transformative. Embrace the challenges ahead as opportunities to leave a lasting impact on your organization, your industry, and the world.

The future is not something to be feared, it is something to be shaped. As a future proof leader, you have the power to guide your organization through uncertainty and into a realm of boundless potential. **Lead boldly, think expansively, and act decisively. The future begins with you.**

As a final leadership discipline, once a year reflects on your own growth arc. **Which signals did you act on early? What risks did you avoid or delay? Which decisions made your team more adaptable?** Use this feedback loop as your personal performance system, not just your team's.

Chapter 16

IT Leadership as a Catalyst for Transformation

Transformation defines the future of successful IT leadership. It is more than adapting to change; it is about initiating it, shaping it, and ensuring it delivers meaningful and lasting progress. The role of an IT leader has evolved beyond managing technology to driving innovation, shaping organizational culture, and crafting strategies that redefine market positioning. As a catalyst for transformation, your influence stretches far beyond the IT department. You are the driving force that aligns technology, people, and purpose in pursuit of creating a more agile, resilient, and future-shaped organization.

This chapter explores how you can embrace the role of a transformational leader, equipping you with the mindset, strategies, and tools to foster progress and inspire those around you while operationalizing transformation as a repeatable discipline, not a one-time campaign.

Chapter 16

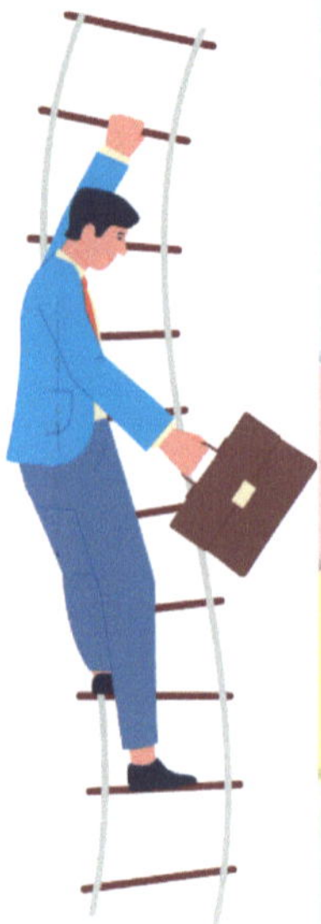

IT LEADERSHIP AS A CATALYST FOR TRANSFORMATION

- **Lead change through vision and innovation.**
- **Cultivate a transformational mindset**
- **Empower teams to drive progress.**
- **Build agility into organizational culture.**

The Transformational Mindset: A New Way of Thinking

Transformation begins within. Before you can inspire others, you must develop a mindset that views challenges as opportunities and limitations as starting points for innovation. A transformational mindset requires vision, adaptability, and resilience, all of which set the tone for your team and the organization.

Cultivating this mindset means consistently looking beyond the immediate challenges to anticipate future trends. Commit yourself to continuous learning by staying informed about emerging technologies, evolving consumer behaviors, and shifting market dynamics. Explore areas outside your traditional expertise, as this broadens your perspective and enables you to connect the dots in new and creative ways.

Resilience is a key component of this mindset. Transformation is rarely linear, and setbacks are inevitable. Instead of fearing failure, view it as a steppingstone toward improvement. Encourage experimentation within your organization and lead by example by embracing calculated risks. The ability to persevere through uncertainty while remaining focused on long term objectives is what separates transformational leaders from the rest.

Build this mindset into practice: Conduct quarterly **Disruption Drills** with your leadership team. Introduce a hypothetical scenario, like a market shift, competitor launch, or regulatory change, and ask: **how would we pivot? Where would we fail fast, and what capabilities would we wish we had already developed?** Use these drills to identify leadership gaps and investment blind spots.

Creating a Vision That Drives Transformation

Transformation without a clear vision is chaos. A compelling vision is the foundation of meaningful change. It provides direction, inspires action, and serves as a unifying force that aligns diverse stakeholders around shared goals.

Your vision should be rooted in a deep understanding of your organization's current challenges and future opportunities. Engage with teams across the organization to gather insights about inefficiencies, pain points, and areas ripe for innovation. Use this knowledge to craft a vision that resonates with employees, stakeholders, and customers alike.

Make your vision relatable and actionable. Rather than presenting it as an abstract ideal, illustrate its practical implications through stories, examples, and scenarios. Show how transformation will enhance customer experiences, streamline processes, or open new revenue streams. The more tangible your vision, the more likely people will rally behind it.

Communicating your vision effectively is equally important. Use every opportunity, meetings, presentations, or informal conversations to reinforce your message. Ensure that your vision remains adaptable to changing circumstances while staying true to its core purpose.

Create a Transformation Narrative Framework: For every initiative, craft a 3-part storyline, **What's broken? Why now? What will change for the people who matter?** Make sure this narrative shows up in all communications, not just strategy decks.

Turn the vision into design criteria for decisions: when selecting vendors, launching programs, or setting OKRs, ask **does this bring us closer to our future state?** If not, revise.

Empowering Teams to Champion Transformation

No leader can drive transformation alone. Your ability to empower others is critical to the success of any change initiative. This involves fostering a culture where creativity, collaboration, and innovation thrive.

Start by creating an environment that encourages experimentation. Empower your teams to propose and test new ideas without fear of failure. When people feel safe to take risks, they are more likely to challenge the status quo and develop innovative solutions.

Invest in professional development to ensure your teams are equipped with the skills and tools they need to succeed. Provide opportunities for training, mentorship, and cross functional collaboration, all of which contribute to a more knowledgeable and adaptable workforce.

Recognize and celebrate contributions to transformational efforts. Publicly acknowledge achievements, whether they come from individuals, teams, or departments. When people feel valued for their work, they are motivated to continue pushing boundaries and striving for excellence.

Establish **Innovation Fellowships** short-term rotations where team members can lead or co-lead high-visibility transformation projects outside their core roles. Use these as both development and discovery tools.

Embed transformation metrics into performance reviews: **Has this person contributed to experimentation, change acceleration, or systemic improvement?** Make transformation part of the career path.

Building Agility into the Organization

Transformation is not a single event but an ongoing process. To sustain it, your organization must be agile enough to respond to new challenges and opportunities as they arise.

Agility begins with creating processes and systems that are flexible and scalable. Within IT, this might mean adopting cloud solutions, modular architectures, or DevOps methodologies. These approaches enable teams to deploy changes quickly and efficiently, ensuring that the organization remains responsive to external shifts.

Agility should extend beyond IT to include the entire organization. Encourage leaders across all departments to adopt principles such as decentralized decision making, iterative development, and continuous feedback loops. These practices enable faster problem solving and more effective collaboration across the enterprise.

Fostering a culture of adaptability is equally important. Encourage open dialogue about lessons learned and areas for improvement. When employees feel their feedback is valued, they are more likely to embrace change and contribute to the organization's transformation journey.

Create a cross-functional **Agility Council,** a rotating body of team leaders and change agents responsible for testing faster decision frameworks, surfacing roadblocks, and piloting nontraditional governance models.

Apply Agile thinking to operational change, not just to product. Run retrospectives on failed policies or underperforming partnerships. Iterate them like you would a prototype.

Leveraging Technology as a Transformation Driver

Technology is the backbone of modern transformation, but it is only effective when implemented strategically. Your role as an IT leader is to ensure that technology aligns with organizational goals and delivers measurable impact.

Start by conducting a comprehensive assessment of your current technology landscape. Identify outdated systems, inefficiencies, and gaps that hinder progress. Collaborate with cross functional teams to determine which technologies have the potential to drive meaningful change.

Prioritize technologies that promote integration, efficiency, and scalability. For example, artificial intelligence can provide data driven insights that enhance decision making, while IoT devices can optimize operations through real time monitoring and analytics. Ensure that every technological investment supports the organization's strategic objectives and enhances its ability to compete in the market.

Successful technology adoption requires effective change management. Provide thorough training and support to help employees embrace new tools and workflows. When teams feel confident in their ability to use technology effectively, they are more likely to embrace transformation initiatives wholeheartedly.

Establish a Transformation Tech Radar, track technologies by readiness, value, and team adoption friction. Review monthly. Use it to prioritize rollouts based on strategic fit and user enablement.

For every major tech investment, launch a **Behavior Shift Sprint**: focus not just on deployment, but on shifting the habits, routines, and workflows the technology is meant to transform.

Overcoming Resistance to Change

Resistance is an inevitable part of transformation, but it does not have to derail your efforts. Addressing concerns proactively and fostering trust, you can turn skeptics into advocates.

Start by communicating openly about the reasons for change. Explain how transformation aligns with the organization's goals and benefits employees, customers, and stakeholders. Transparency builds credibility and reduces uncertainty.

Empathy is another powerful tool for overcoming resistance. Take the time to understand why people may be hesitant about change. Whether it's fear of failure, job security concerns, or discomfort with new technologies, acknowledging these feelings and providing support can ease the transition.

Involve employees in the transformation process to create a sense of ownership and engagement. When people feel their voices are heard and their contributions matter, they are more likely to support the changes being implemented.

Run pre-launch **Pushback Mapping** workshops: Ask frontline teams to list all the reasons change may fail, politically, technically, or emotionally. Use their feedback to preempt resistance and improve your plan.

Track transformation sentiment like a product: use pulse surveys to monitor buy-in, friction, and trust before, during, and after rollout. Respond visibly.

Measuring Impact and Sustaining Momentum

Transformation must produce measurable results to be considered successful. Define clear metrics to track progress, such as operational efficiencies achieved, customer satisfaction improvements, or revenue growth attributed to new initiatives.

Regularly review these metrics to assess the effectiveness of your strategies. Use data insights to identify areas where adjustments are needed and opportunities for further innovation. Transformation is an iterative process, and continuous improvement is essential to maintaining momentum.

Celebrate milestones and reflect on lessons learned along the way. Recognizing progress reinforces the value of transformation and motivates teams to continue striving for excellence.

Build a Transformation Value Tracker that ties initiatives to outcomes, cost avoided, speed gained, satisfaction improved. Show how transformation directly links to value.

Normalize **Post-Mortem Learning Loops** after major initiatives, capture surprises, missed bets, and unexpected wins. Feed these back into your transformation playbook.

Inspiring a Legacy of Transformation

Your role as an IT leader extends beyond initiating change. It is about creating a culture where transformation becomes a continuous pursuit, embedded in the fabric of the organization.

Inspire your teams to think boldly and act decisively. Demonstrate the value of innovation and adaptability through your actions and leadership style. Empowering others to embrace transformation, you create a ripple effect that extends far beyond your immediate influence.

The legacy of a transformational IT leader is one of lasting impacts. It is reflected in the resilience, creativity, and agility of the organization you help build. Embrace your role as a catalyst for change, and you will leave behind a foundation of progress and possibility that shapes the future for years to come.

Document your transformation **firsts** first pilot, first resistance, first breakthrough, and share them with rising leaders. Turn your journey into a leadership accelerator.

Close each major cycle of change with a reflection: **What did we shift? What did we learn? What's next?** Make your transformation story visible, and unfinished.

Part 6

Innovating for the Future (Chapters 17 - 18)

Part 6 emphasizes the importance of fostering innovation and leveraging transformative technologies to stay ahead in the ever-evolving IT landscape.

In **Chapter 17**, Cultivating a Culture of Continuous Innovation, leaders explore strategies to embed creativity and adaptability into the organizational culture. This chapter provides actionable guidance on encouraging experimentation, institutionalizing innovation processes, and recognizing contributions that drive progress. It underscores the role of leadership in creating an environment where bold ideas can flourish and fuel sustained growth.

Chapter 18, AI Leadership: Harnessing the Future Today, focuses on the groundbreaking potential of artificial intelligence to revolutionize industries and drive transformative change. IT leaders learn how to integrate AI into their strategies, ensure responsible and ethical implementation, and position their organizations as pioneers in the AI driven future. This chapter combines visionary insights with practical steps to harness AI's full potential while mitigating risks.

Together, these chapters prepare IT leaders to innovate boldly, adapt swiftly, and lead their organizations into a future defined by creativity and technological advancement.

Chapter 17

Storytelling as a Leadership Tool in IT

The complex and data driven world of IT leadership often relies on metrics, logic, and analytics to make decisions and solve problems. Yet there is a powerful tool that can elevate the effectiveness of IT leaders beyond mere numbers and technical jargon: storytelling. The ability to tell a compelling story bridges the gap between abstract technological concepts and human understanding, fostering connection, alignment, and action.

Storytelling has the unique capacity to convey purpose, simplify complexity, and inspire teams. In IT leadership, it transforms data into meaning, processes into narratives, and strategies into shared visions. This chapter explores the transformative power of storytelling, its vital role in IT leadership, and how you can master this skill to elevate your impact.

The Role of Storytelling in IT Leadership

Storytelling succeeds because it is deeply ingrained in human nature. While data appeals to logic, stories resonate with emotion, creating a lasting impression that numbers alone cannot achieve. As an IT leader, much of your work involves systems, processes, and innovations that are abstract or technical. Storytelling helps translate these complexities into relatable narratives that inspire and inform.

For instance, explaining a cybersecurity initiative by detailing its technical components might lose your audience in jargon. However, sharing the story of a similar organization that suffered a significant breach and the consequences they faced makes the same case in a relatable, memorable way. Stories bring clarity and urgency to abstract concepts, ensuring your audience not only understands but also cares.

IT leaders must engage with diverse audiences, from technical teams to executive stakeholders, each with different priorities and levels of expertise. Storytelling allows you to meet each group where they are, presenting information in a way that aligns with their interests and motivates them to act.

Embed this as a strategic practice: Build a **Story Matrix** for each stakeholder group, Executives, Teams, Partners, and Customers. For each, outline 3 common concerns and align them with pre-crafted stories that reinforce clarity, urgency, or value. Maintain this matrix as a living part of your leadership toolkit.

Why Storytelling Matters in IT

Storytelling is more than a communication technique; it is a leadership strategy. It is a tool for forging connections, building trust, and aligning teams toward a shared vision. It creates context for decision making, turning abstract ideas into concrete action points.

As an IT leader, storytelling enables you to:

- ✓ Simplify complex concepts by framing them within relatable and engaging narratives.
- ✓ Inspire change by connecting transformation initiatives to a broader purpose that resonates with your audience.
- ✓ Build trust by sharing authentic experiences that demonstrate vulnerability, wisdom, and empathy.
- ✓ Persuade stakeholders to support initiatives by illustrating their potential impact through real life scenarios.

Through storytelling, you can foster a culture of innovation and alignment, motivate your team and create buy-in from stakeholders. A well-told story captures the imagination, engages the heart, and solidifies commitment.

Operationalize this by embedding **Story Slides** into every major initiative deck. Before data and outcomes, open with a brief, relevant narrative that answers: **Why now? Who's affected? What's the vision?** Use this structure to consistently anchor logic in meaning.

Crafting Stories with Purpose

Great stories are not random anecdotes; they are intentional narratives that serve a clear purpose. Crafting an effective story requires you to define your objective, understand your audience, and weave a compelling narrative that aligns with your goals.

Begin by asking yourself: What do I want my audience to think, feel, or do after hearing this story? Your objective could be to inspire your team to embrace a challenging transformation, convince executives to approve a new project, or reassure stakeholders during a crisis. The purpose of your story will shape its structure and tone.

Every compelling story includes a relatable protagonist, a challenge or conflict, and a resolution. In IT, your protagonist might be a team member solving a complex issue, a customer benefiting from a solution, or even the organization itself overcoming an obstacle. The challenge should create tension, engage your audience and emphasize the stakes. Finally, the resolution provides a satisfying conclusion that reinforces the lesson or action you want to communicate.

The best stories are those that reflect authenticity. Draw from real life experiences within your career or organization to ensure your stories feel genuine and relatable.

Develop a 3-Part **Leadership Story Arc** template: (1) The tension, a real challenge you faced; (2) The turning point, what insight, action, or mindset changed the course; (3) The takeaway, what you want your audience to apply. Use this arc to design every strategic story you deliver.

Institutionalize it: Train leaders in your org to use this arc during town halls, onboarding, and product launches to create alignment and continuity of message.

Tailoring Stories to Your Audience

An impactful story resonates with its audience, aligning with their concerns, goals, and perspectives. To tailor your story effectively, start by understanding your audience's priorities.

When addressing your team, emphasize their role in the larger mission. Share stories that highlight their contributions, celebrate successes, and demonstrate how their work impacts the organization and its customers. For example, recount how their effort in resolving a critical system outage ensured uninterrupted services for thousands of users, positioning them as heroes of the narrative.

When speaking to executives, focus on strategic outcomes. Your stories should illustrate how an initiative aligns with organizational goals, generates ROI, or mitigates risks. For instance, if pitching a data analytics platform, share a story of a similar implementation that unlocked valuable insights, drove decision making, and increased profitability.

When engaging with customers, highlight trust, value, and results. Share stories of how your solutions addressed similar challenges for other clients, showcasing your organization's reliability and innovation. This reinforces your credibility and deepens customer loyalty.

Tailoring your stories ensures they resonate on a personal and professional level, maximizing their impact.

Systematize this through a **Story Persona Map** for key roles in your ecosystem, internal teams, investors, users, regulators. For each, identify their pain points, values, and success definitions. Use this to select and shape stories for maximum impact.

Building Storytelling Skills

Storytelling is not an innate talent but a skill that can be developed through intentional practice and reflection. To become a more effective storyteller, you must hone your ability to craft, deliver, and refine your narratives.

Reflect on pivotal moments in your career. Consider the challenges you've faced, the lessons you've learned, and the successes you've achieved. These experiences often form the foundation of the most compelling stories because they are authentic and relatable.

Create a library of stories that align with your leadership objectives. Over time, you will accumulate examples to draw from, whether you need to motivate your team, persuade stakeholders, or explain a complex concept.

Practice regularly in everyday interactions. Look for opportunities to incorporate storytelling into meetings, presentations, and informal conversations. The more you practice, the more natural it will become to weave stories into your leadership style.

Seek feedback after sharing a story. Ask trusted colleagues or team members if your narrative resonated, inspired action, or clarified a concept. Use their feedback to refine your storytelling techniques and better understand what works for different audiences.

Build a **Leadership Story Vault** a shared, searchable internal database of high-impact stories categorized by theme: trust, transformation, failure, innovation, resilience. Encourage all leaders to contribute. Use it in planning strategic communications, leadership training, and board updates.

Storytelling in Action

The real impact of storytelling lies in its application to everyday leadership challenges. Consider how storytelling can transform the following scenarios:

- When leading a digital transformation, narrate how a similar initiative enabled another organization to achieve remarkable results. Share the steps they took, the obstacles they overcame, and the benefits they reaped. This makes the transformation feel tangible and achievable for your team.
- During times of resistance to change, share a story of a hesitant group that embraced a new technology and ultimately improved their workflows and morale. Highlight how their journey mirrored the current situation, showing how resistance can turn into progress.
- When celebrating success, tell a story that showcases the contributions of an individual or team. Detail the challenges they faced, their problem-solving approach, and the positive outcomes of their efforts. This not only boosts morale but sets a standard of excellence for others to emulate.

Create a **Story Activation Checklist** for leaders managing major changes. Include prompts like: **What story frames this initiative? Whose journey can illustrate the stakes? How can we spotlight team effort through narrative?** Deploy the checklist in every transformation kickoff.

Taking Storytelling Beyond Words

Stories don't have to rely solely on spoken or written words. Visuals, demonstrations, and data can amplify the impact of your narratives. A compelling chart, image, or video can bring your story to life, making abstract concepts tangible and engaging.

For example, when sharing a story about reducing downtime through system upgrades, it includes a graph that shows the before and after performance metrics. The visual reinforces the story's message, ensuring it resonates more deeply with your audience.

Build **Story + Signal** presentations, every story includes a paired visual indicator: performance metric, timeline, or impact delta. Use this model to combine narrative with evidence in every executive or operational presentation.

Fostering a Storytelling Culture

The power of storytelling should not rest solely with you as the leader. Encouraging your team to share their stories, you create a culture of connection, collaboration, and innovation.

Highlight team members' stories during meetings, presentations, or internal communications. Celebrate their successes, lessons, and insights, creating a shared narrative that unites the organization around its values and goals.

Provide opportunities for employees to develop their storytelling skills. Host workshops, invite guest speakers, or create platforms where team members can share their experiences. A storytelling culture fosters engagement, trust, and alignment, amplifying the collective power of your organization.

Institutionalize this with a monthly **Leadership Story Circle** an open session where leaders across teams share real experiences from the field tied to strategy, resilience, or transformation. Document key learnings and publish highlights across the org.

The Legacy of Storytelling in IT Leadership

Storytelling is more than a communication tool; it is a leadership strategy that transforms how people connect, understand, and act. As an IT leader, your stories bridge the gap between technology and humanity, creating meaning from complexity and inspiring those around you to achieve greatness.

Every challenge you face, every innovation you lead, and every success you celebrate is a story waiting to be told. Use these moments to build trust, inspire action, and reinforce your vision. When you lead with stories, you leave behind a legacy that extends beyond systems and processes, it touches the hearts and minds of the people you lead.

The power of storytelling is now in your hands. Use it to create connections, drive transformation, and lead with impact. Your audience is ready, and your story is waiting to be told.

Make this your signature: Capture and publish your **Leadership Storybook** as part of succession planning. Share the pivotal stories that shaped your leadership journey. Make them accessible to the next generation. Let them learn through the power of your lived experience.

The power of storytelling is now in your hands. Use it to create connections, drive transformation, and lead with impact. Your audience is ready, and your story is waiting to be told.

Chapter 18

Cultivating a Culture of Continuous Innovation

Innovation thrives where curiosity meets action, where creativity is not a luxury but a daily practice, and where every member of an organization feels empowered to challenge the ordinary. Building a culture of continuous innovation means embedding these ideals into the fabric of your leadership and organization. This is not the result of spontaneous flashes of genius but the product of intentional strategies, consistent leadership, and a deep commitment to progress.

Creating an environment where innovation flourishes require a deliberate blend of vision, structure, and support. It demands a leadership approach that encourages risk taking, celebrates learning, and builds systems that transform ideas into impact.

Vision as the Catalyst for Innovation

Continuous innovation begins with a vision that inspires action and aligns efforts. Without a shared purpose, even the most talented teams can drift into silos, focusing on isolated efforts that lack cohesion. Your vision should answer a fundamental question: What are we striving to achieve through innovation, and why does it matter?

This vision should not be confined to organizational objectives alone. It must resonate on a personal level, connecting everyone's contributions to a larger purpose. For example, if your organization is committed to improving healthcare access through technology, illustrate how the work done today could save lives tomorrow. When employees understand the real-world implications of their efforts, they approach challenges with a sense of purpose and urgency.

Make your vision a living narrative, something that evolves and adapts with your organization. Reinforce it regularly, using examples, stories, and successes to keep it vivid and relatable. A vision that is understood and embraced becomes the foundation upon which innovation is built.

Operationalize this by creating a **Vision Alignment Loop**. Every 60 days, review one key innovation effort and re-map how it supports your evolving mission. Ask: **Is this initiative solving what matters most now? Where do we need to pivot to stay relevant and bold?** Keep this loop visible in strategic planning cycles.

Building an Environment of Trust and Exploration

Innovation thrives in an atmosphere of trust, where individuals feel safe to share ideas, take risks, and embrace experimentation. The fear of failure or judgment stifles creativity, reducing the willingness to propose unconventional solutions or pursue bold initiatives.

To foster this environment, leaders must create a culture where mistakes are seen as opportunities for learning. When an experiment doesn't yield the desired results, shift the focus from assigning blame to understanding lessons. This approach not only removes the fear of failure but also reinforces the importance of persistence and resilience.

Encourage open dialogue within your teams. Actively seek input from all levels of the organization, ensuring that every voice is valued. When people feel heard, they are more likely to contribute their ideas and insights, even if those ideas challenge the status quo.

Celebrate curiosity and exploration. Acknowledge the courage it takes to propose new approaches or question established practices. Recognition and encouragement fuel a culture where innovation becomes second nature.

Embed a **No-Blame Debrief** routine. After any failed or paused project, hold a 20-minute team reflection focused only on what was tried, what was learned, and what should be attempted differently next time. Publish one lesson from these in a shared internal channel monthly.

Build a **Greenlight Card** process, where any employee can anonymously pitch an idea to be reviewed by a rotating innovation panel. Select one every quarter for experimentation funding or prototyping.

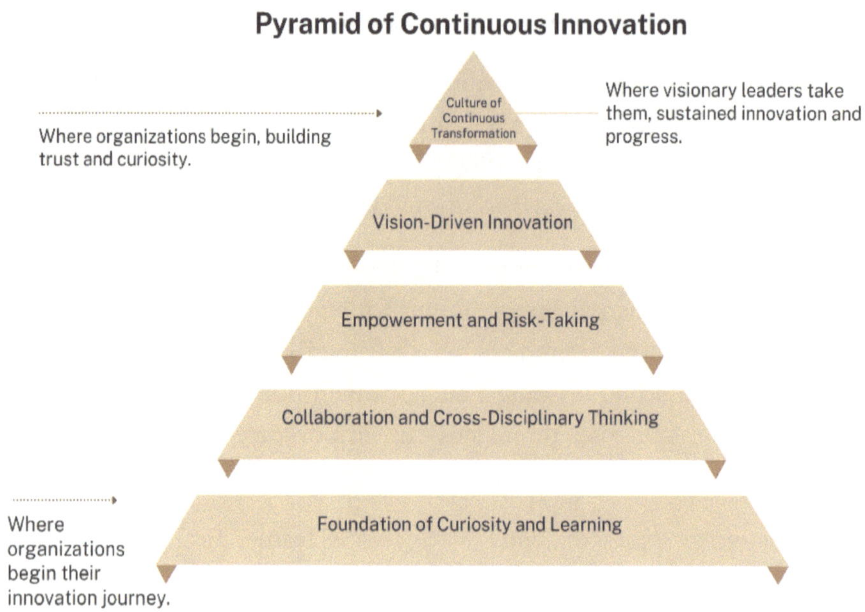

Encouraging Collaboration Across Boundaries

Collaboration is a powerful driver of innovation, as diverse perspectives often lead to breakthrough ideas. Bringing together individuals with different skills, backgrounds, and experiences, you create fertile ground for creativity and problem solving.

Start by breaking down silos that separate departments or teams. Establish cross functional projects where marketing collaborates with IT, finance partners with operations, and product teams work alongside customer service. When people work together across disciplines, they gain new insights and identify solutions that might not emerge within a single domain.

Facilitate regular forums for knowledge sharing, such as workshops, brainstorming sessions, or team retreats. Encourage participants to discuss challenges, share successes, and explore opportunities together. These gatherings not only spark new ideas but also build relationships that strengthen collaboration over time.

Provide tools and platforms that make collaboration seamless, whether through project management software, communication channels, or shared workspaces. The easier it is for people to work together, the more likely they are to do so effectively.

Formalize this through **Innovation Pairing**. Once per quarter, assign pairs of employees from different departments to shadow each other and co-design a solution to a shared friction point. Present findings to leadership with no pressure to implement, just to surface cross-boundary insight.

Structuring Systems for Experimentation

For innovation to be sustainable, it must be supported by systems that encourage and reward experimentation. Without the right processes and resources, even the best ideas can falter before they have a chance to grow.

Dedicate time and resources to innovation. Consider implementing dedicated innovation sprints or setting aside time for employees to explore creative projects outside their regular responsibilities. These initiatives signal that innovation is a priority, not an afterthought.

Provide the tools and infrastructure needed to experiment effectively. For example, create a sandbox environment where IT teams can test new solutions without impacting live systems. Allocate budgets for pilot programs, allowing teams to prototype and refine their ideas before scaling them.

Develop clear pathways for scaling successful experiments. When an idea demonstrates potential, it has processes in place to move it from concept to implementation quickly and efficiently. This ensures that innovation doesn't get stuck in endless planning stages.

Implement a **Stage-Gate Innovation Pipeline** with three visible tiers: Explore (ideas in play), Validate (pilots in motion), Expand (proven and scaling). Review this monthly pipeline with input from across teams.

Introduce a **24-Hour Hackpad** process: Any employee can submit a lightweight concept on a rolling basis. A small panel evaluates ideas within 24 hours, offering either next steps or coaching. This maintains speed and responsiveness.

Inspiration Through Recognition and Reward

A culture of continuous innovation requires reinforcement. Recognizing and rewarding innovation sends a powerful message that creative thinking and problem solving are valued throughout the organization.

Celebrate both successes and efforts. While achieving measurable outcomes is important, the process of exploring new ideas and testing possibilities is equally deserving of recognition. Highlight these contributions in meetings, newsletters, or companywide communications to inspire others.

Establish formal and informal incentives for innovation. These might include awards for the most impactful ideas, opportunities for professional development, or even small gestures like personalized thank you notes. The key is to show that innovation is appreciated and encouraged at all levels.

Create an **Innovator of the Month** spotlight, not based solely on results but on boldness, collaboration, and learning mindset. Share these stories internally as mini-case studies to normalize risk-taking and experimentation.

Use micro-rewards: a handwritten note, a spontaneous recognition in a team chat, or access to a unique learning opportunity. Small signals drive big cultural outcomes.

Harnessing Data and Insights for Continuous Improvement

Innovation is most effective when it is informed by data. Analyzing trends, gathering feedback, and measuring outcomes, you can refine your efforts and ensure they deliver meaningful results.

Monitor metrics that matter to your organization, whether they relate to customer satisfaction, operational efficiency, or employee engagement. Use this data to identify gaps, uncover opportunities, and track the impact of innovative initiatives.

Create feedback loops with both employees and customers. Employees often have firsthand knowledge of inefficiencies or challenges, while customers provide valuable insights into how your products or services are perceived. Regularly seek their input and use it to guide your innovation strategy.

Data driven decision making ensures that innovation is not only creative but also strategic. It aligns your efforts with organizational goals and maximizes their impact.

Install a **Continuous Listening Loop** quarterly internal surveys and semiannual customer interviews specifically designed to surface innovation blockers and unmet needs. Feed this input into a central backlog reviewed by innovation leads.

Pair every pilot project with a real-time data dashboard. Define 2-3 outcome metrics before starting, then share performance updates transparently across teams, even if results are mixed.

Leading by Example

As a leader, your actions set the tone for your organization's culture. To cultivate continuous innovation, you must model the behaviors and attitudes you want to see in your team.

Demonstrate curiosity and a willingness to explore new ideas. Show enthusiasm for learning about emerging trends, technologies, and methodologies. When your team sees you actively engaging with innovation, they'll be inspired to do the same.

Be transparent about your own experiments and initiatives, sharing both successes and setbacks. This vulnerability reinforces the idea that innovation involves risk and that even leaders are not immune to challenges.

Support your team by removing obstacles that hinder creativity, whether they are bureaucratic processes, resource limitations, or outdated mindsets. Advocate for the changes needed to create an environment where innovation can thrive.

Adopt a **Show Your Drafts** principle, share early ideas, half-baked concepts, and incomplete prototypes with your team before polishing. This reduces perfectionism and models comfort with ambiguity.

Regularly ask one question in your one-on-ones: **What's something you've always wanted to try here, but haven't yet?** Capture themes and act on the patterns that emerge.

Embracing the Journey of Innovation

Innovation is not a destination to be reached but a journey to be embraced. It requires a mindset that views change as an opportunity rather than a threat, a commitment to continuous growth, and a willingness to adapt as circumstances evolve.

Cultivating a culture of continuous innovation, you position your organization to lead rather than follow, to adapt rather than react, and to thrive in the face of uncertainty. This culture is not built overnight, but with consistent effort, it becomes a defining characteristic that sets your organization apart.

Take the first step by articulating a clear vision, building trust, encouraging collaboration, and providing the tools and support needed for experimentation. Lead with curiosity, embrace challenges, and celebrate the progress made along the way.

The future belongs to organizations that prioritize creativity, adaptability, and progress. Fostering a culture of continuous innovation, you're not just preparing your organization for the future, you're shaping it.

Final note: Set a **Year One Innovation Audit** for your team. Twelve months from now, reflect together: **What risks did we take? What friction did we remove? What bold idea surprised us?** Let this retrospective become your new roadmap.

Part 7

Legacy and Impact

(Chapters 19 - 21)

Part 7 delves into the enduring influence of IT leadership and the profound impact leaders can leave on their organizations, industries, and communities. This part encapsulates the essence of mentorship, reflection, and legacy building, offering a holistic view of leadership's ultimate purpose and reach.

Chapter 19, Mentorship in IT Leadership, emphasizes the importance of guiding the next generation of leaders. It provides actionable strategies for fostering talent, sharing knowledge, and building a culture of learning and inclusivity. This chapter showcases how mentorship can multiply a leader's impact by empowering others to succeed.

Chapter 20, From Visionary to Legend, reflects on the leadership journey, summarizing key principles from the book. It inspires readers to lead boldly, embrace their unique strengths, and turn their vision into a lasting impact, both professionally and personally.

Chapter 21, Leaving a Legacy of Transformative IT Leadership, focuses on creating a legacy that endures beyond tenure. It explores strategies for embedding values, fostering innovation, and shaping a future oriented culture.

Together, these chapters equip IT leaders to craft a legacy defined by empowerment, innovation, and meaningful change.

Chapter 19

Mentorship in IT Leadership

Mentorship stands as one of the most profound responsibilities of IT leadership. It serves as a channel to cultivate talent, amplify organizational capabilities, and build a legacy that extends beyond immediate achievements. Effective mentorship in IT is a deliberate effort to guide, empower, and inspire others to grow, both professionally and personally. For leaders, it offers an opportunity to foster innovation, nurture resilience, and shape the next generation of visionaries.

This chapter explores mentorship as a transformative leadership tool. Understanding its principles, practicing intentional approaches, and embedding it within organizational culture, you can become a mentor who shapes not only careers but the very future of IT leadership.

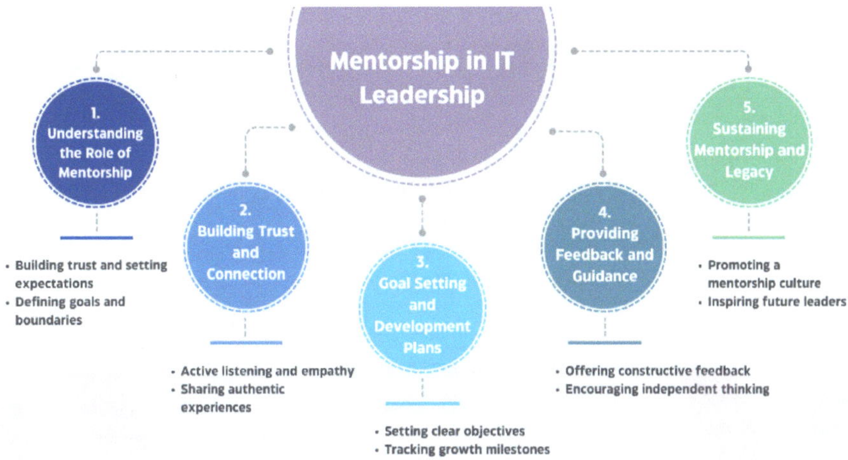

Understanding the Impact of Mentorship

Mentorship is more than offering advice or guidance. It is an investment in the growth of others that simultaneously strengthens the mentor's leadership capabilities. When leaders mentor effectively, they contribute to the development of skills, confidence, and leadership qualities in their mentees while refining their communication, empathy, and strategic thinking.

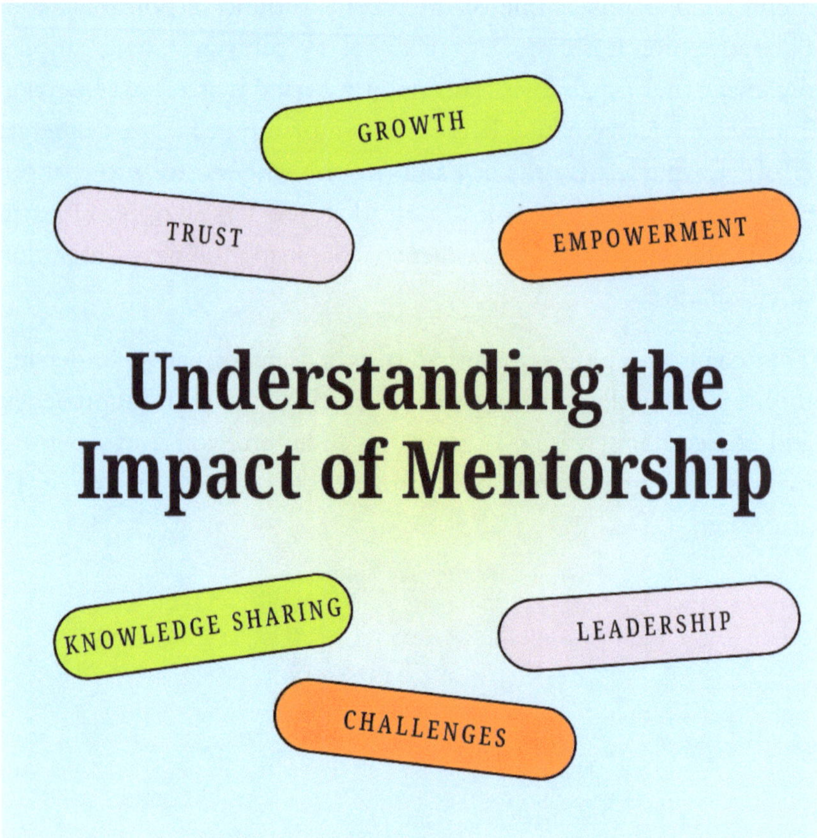

In IT, where innovation drives success and challenges and evolve rapidly, mentorship ensures continuity of expertise and adaptability. Through mentorship, IT leaders create environments where knowledge is shared, collaboration is enhanced, and individuals feel empowered to step into leadership roles themselves.

Effective mentorship builds lasting trust and creates a ripple effect. The lessons learned by one mentee are often passed on to others, amplifying the mentor's influence across teams, departments, and even industries.

As a mentor, start by mapping your **mentorship sphere of impact**. Identify three individuals in your ecosystem, within or beyond your team, whose growth could multiply organizationally value if nurtured. Treat this as your mentorship portfolio and revisit progress quarterly.

Document your mentorship stories, real successes and setbacks, and share them with peers. This transparency normalizes the process and helps build a collective mentorship mindset across departments.

Defining a Purposeful Approach to Mentorship

To be an effective mentor, you must approach the role with intention and clarity. The foundation of mentorship lies in understanding its purpose. It is not about solving every problem for your mentee but about equipping them with the tools and perspectives to solve problems on their own.

The process begins by setting clear expectations. Discuss the goals and boundaries of the mentorship relationship early on. Are you helping your mentee navigate career transitions, develop technical expertise, or enhance leadership skills? Defining the purpose provides direction and ensures that both parties are aligned.

Equally important is understanding the individual you are mentoring. Take the time to learn their strengths, weaknesses, aspirations, and challenges. This insight enables you to tailor your guidance, ensuring it resonates and drives meaningful progress.

Create a "**Mentorship Blueprint**" document together. Outline objectives, success indicators, meeting cadence, and expected learning outcomes. This written commitment turns mentorship from informal dialogue into a structured development agreement.

Revisit this blueprint every 90 days to recalibrate goals, what's working, what's changed, what new skills should be added. Keep it dynamic and aligned with your mentee's evolution.

Establishing Trust as the Foundation

Trust is the bedrock of a successful mentorship relationship. Without it, mentees may hesitate to share their challenges, and the guidance offered may lack authenticity. Establishing trust requires consistency, empathy, and vulnerability.

Begin by being fully present during your interactions. Show your mentee that their growth is a priority by listening actively, asking thoughtful questions, and responding with sincerity. Share your own experiences, including the setbacks and lessons that shaped you as a leader. This transparency not only builds trust but also demonstrates that challenges are part of growth.

Respect your mentee's autonomy. While your role is to guide and support, it is equally important to encourage independent decision making. Allow them to take ownership of their journey, even if it means navigating setbacks along the way.

Implement **Trust First Conversations**. In early sessions, avoid technical topics and instead explore shared experiences, personal motivations, and long-term aspirations. Human connection forms before professional direction.

Adopt the 70/30 listening rule, listen 70% of the time, speak 30%. This creates psychological safety and ensures mentorship feels like empowerment, not evaluation.

Goal Setting as a Mentorship Strategy

Goals provide structure and focus to mentorship, ensuring that progress is tangible and measurable. As a mentor, you play a pivotal role in helping your mentee identify and refine their objectives.

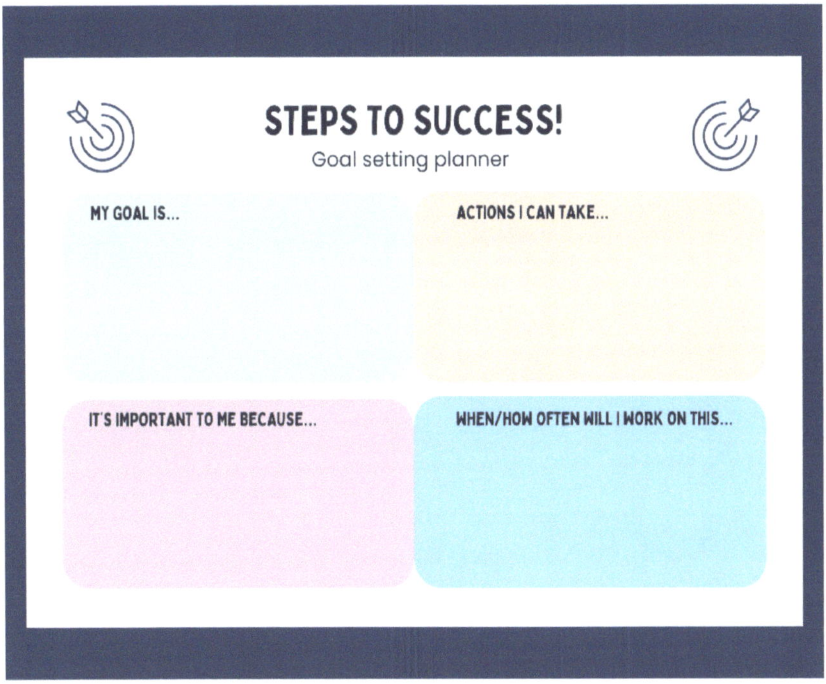

Work collaboratively to define clear, actionable goals. Encourage your mentee to articulate their aspirations and break them into manageable milestones. For instance, if a mentee aspires to lead a complex IT project, their milestones might include mastering project management frameworks, build cross functional relationships, and developing strategic planning skills.

Revisit these goals regularly to assess progress and adapt as needed. Celebrate achievements, no matter how small, to build confidence and momentum. When setbacks occur, use them as opportunities to reflect and recalibrate. The process of goal setting and review reinforces accountability and demonstrates the value of sustained effort.

Use a **SMART Growth Map** for every mentorship engagement: Specific, Measurable, Achievable, Relevant, Time-bound. Plot milestones on a visual tracker visible to both mentor and mentee. Visualizing progress reinforces motivation.

Schedule **Checkpoint Conversations** every month. Ask three questions: **What did you attempt this month? What did you learn? What will you try next?** Keep feedback short, candid, and tied to growth outcomes.

Providing Guidance Through Active Engagement

Effective mentorship is an active and dynamic process. It requires you to balance offering guidance with encouraging independence. Rather than providing solutions outright, ask questions that help your mentee reflect and find their own answers. This approach fosters critical thinking and empowers them to tackle challenges with confidence.

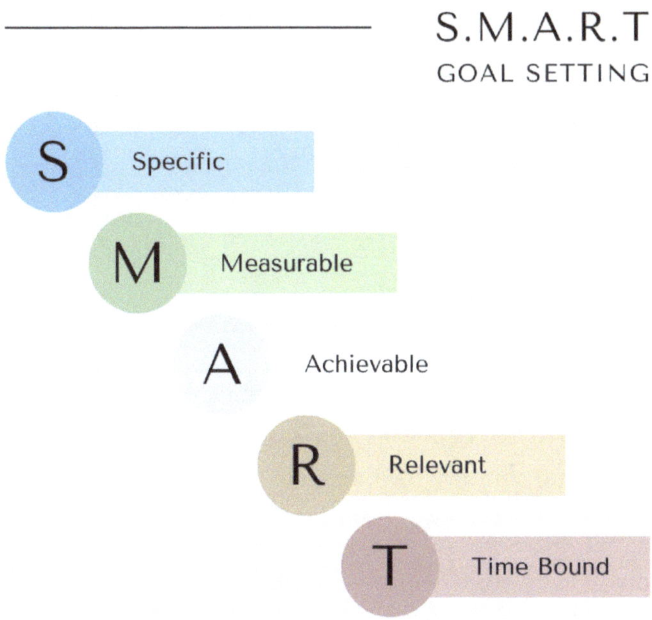

Create opportunities for real-world learning. Assign projects or tasks that stretch your mentee's abilities and expose them to new experiences. For instance, invite them to shadow you during strategic meetings, assign them to lead a smaller project, or involve them in decision-making processes. Ensure these opportunities are **Specific,**

Measurable, Achievable, Relevant, and Time Bound to maximize their learning outcomes. These hands-on experiences accelerate growth and build practical skills.

Feedback is another essential element of mentorship. Offer constructive feedback that is specific, actionable, and supportive. Frame critiques as opportunities for improvement and pair them with encouragement to reinforce your belief in their potential.

Apply the **Reflect-Refine-Repeat** feedback loop. After each task or project, hold a 15-minute discussion focused on three elements: what worked well, what could be refined, and what to repeat next time.

Encourage mentees to keep a **Leadership Learning Journal**. After each mentorship session, they record insights, applied lessons, and next steps. Review it together periodically, it builds self-awareness and progress tracking.

Navigating Challenges Together

Every mentorship journey encounters challenges, whether they stem from misaligned expectations, resistance to feedback, or external pressures. Your ability to navigate these obstacles defines the strength and impact of the relationship.

When challenges arise, approach them with empathy and patience. Encourage open dialogue to understand the root causes and work collaboratively to find solutions. Maintain a focus on growth, emphasizing that challenges are part of the process and provide valuable learning opportunities.

As a mentor, you must also be prepared to adapt your approach. Each mentee is unique, and what works for one individual may not resonate with another. Stay attuned to their needs and adjust your guidance to meet them where they are.

Use a **Mentorship Pulse Check** every quarter, short anonymous feedback from your mentee on what's working and what could improve. Treat it as a partnership, not an evaluation.

Model resilience in real time. When external setbacks occur, share how you process and respond to adversity. Mentees learn far more from witnessing recovery than from hearing about success.

Sustaining Mentorship Beyond Immediate Goals

The true impact of mentorship extends far beyond the achievement of immediate goals. A successful mentorship relationship equips mentees with principles and skills that guide them throughout their careers.

Encourage your mentee to adopt a growth mindset, emphasizing the importance of continuous learning and adaptability. Instill habits such as seeking feedback, setting ambitious goals, and reflecting on experiences. These practices ensure that the lessons of mentorship remain relevant and impactful long after the formal relationship concludes.

Stay connected with your mentee as they progress in their career. Celebrate their successes, offer guidance during transitions, and continue to serve as a trusted advisor. This ongoing connection strengthens the bond you've built and reinforces your role as a lifelong mentor.

Encourage mentees to **pay it forward**. Once they reach a milestone, have them mentor someone new using the same framework. This multiplies the impact of your mentorship across the organization.

Create a **Mentorship Continuum** tracker, a simple visual showing where your past mentees now stand and how they've contributed to mentoring others. It becomes living proof of your leadership legacy.

Building a Culture of Mentorship

Mentorship should not be limited to one-on-one relationships. As a leader, you have the ability to create an organizational culture where mentorship is a shared value and practice.

Promote mentorship at all levels of the organization, encouraging employees to both seek guidance and offer it to others. Develop formal mentorship programs with clear objectives, training, and support. Recognize and reward mentorship efforts to highlight their importance and inspire participation.

A culture of mentorship enhances talent development, strengthens collaboration, and fosters a sense of community within your organization. It ensures that knowledge and expertise are continually passed down, building a resilient and innovative workforce.

Launch a **Mentorship Hub** within your organization, a platform or forum where employees can find mentors, share resources, and exchange insights across departments and geographies.

Institutionalize mentorship through onboarding and leadership programs. Make mentoring hours part of performance metrics, reinforcing that developing others is a measure of leadership success.

Leaving a Legacy Through Mentorship

Mentorship is one of the most enduring legacies a leader can create. Investing in others, you amplify your influence and contribute to a cycle of growth and leadership development. The mentees you guide today will become the mentors of tomorrow, perpetuating the values and principles you instilled.

As you reflect on your journey as a mentor, consider the impact you wish to leave behind**. How will your guidance shape the careers and contributions of those you mentor? How will their successes reflect your own commitment to leadership and growth?**

Affirm the power within your mentees. Encourage them to embrace identities that empower:

I AM loved. I AM worthy. I AM blessed. I AM healed. I AM brave. I AM happy. I AM cool.

These affirmations are not just words, they are foundations of confidence and pillars of self-worth that strengthen your mentee's ability to lead with authenticity and resilience.

Embrace mentorship as a lifelong responsibility and privilege. Through mentorship, you have the opportunity to shape the future of IT leadership, one individual at a time. Start now, with purpose and passion, and watch as your efforts create ripples of transformation that extend far beyond your reach.

Before closing each mentorship chapter, write a short **Legacy Letter** to your mentee, a reflection of what they've achieved, how they've grown, and what you hope they'll carry forward. These letters become milestones of purpose for both of you.

Chapter 20

AI Leadership: Harnessing the Future Today

Artificial Intelligence stands as a transformative force, redefining industries, reshaping economies, and reframing the way we live and work. As a leader in IT, your engagement with AI is no longer optional, it is essential. The challenge before you is not merely to deploy AI but to lead its thoughtful integration into your organization, aligning its vast potential with strategic goals and ethical principles.

AI leadership also demands continuous learning, adaptive thinking, and the courage to challenge traditional norms. You must foster innovation while safeguarding privacy, equity, and accountability. Empowering teams to co-create with AI will be a hallmark of visionary leadership in this era.

AI represents more than technological advancement; it is a paradigm shift that demands a new kind of leadership, one that understands its intricacies, anticipates its implications, and embraces its opportunities with foresight and responsibility.

Unveiling the Transformative Potential of AI

AI's potential lies not only in automating tasks but in revolutionizing processes, decision making, and value creation. From predicting customer behaviors with precision to enabling real time insights for operational efficiency, AI equips organizations to think beyond traditional constraints.

As a leader, your first step is to identify the areas within your organization where AI can deliver the most meaningful impact. This requires assessing processes that can benefit from automation, customer interactions that can be enhanced through personalization, and strategies that can be informed by predictive analytics. Your ability to align AI's capabilities with your organization's broader vision will determine the scale of its transformative effect.

However, this transformation is not without challenges. The deployment of AI requires a balance between ambition and pragmatism. Leaders must manage expectations, emphasizing that while AI can unlock significant potential, its true value comes from careful integration and continuous optimization.

Start with a **Transformation Discovery Sprint**. Over four weeks, map every workflow where time-consuming manual judgment occurs, claims processing, scheduling, forecasting, citizen-service triage, and rate each by automation potential, ethical sensitivity, and measurable value. This becomes your AI impact roadmap.

Build a cross-industry lens. Study how healthcare uses AI for diagnostics, how finance applies it to fraud detection, and how manufacturing automates quality control. Then translate insights into your context through "adjacency workshops" that ask: **What would this look like in our world?**

HOW TO LEAD WITH AI IN 5 STRATEGIC STEPS

The Responsibilities of an AI Leader

To lead AI initiatives effectively, you must cultivate both a deep understanding of its technical foundations and an acute awareness of its human impact. AI leadership begins with education, demystifying its mechanics and helping your team and stakeholders grasp its capabilities, limitations, and implications. Fostering a culture of understanding, you enable your organization to approach AI with confidence rather than apprehension.

Ethics is a central pillar of AI leadership. Issues such as algorithmic bias, data privacy, and the societal effects of automation require you to navigate complex questions with clarity and integrity. Setting and enforcing ethical guidelines ensures that AI is developed and deployed responsibly, balancing innovation with accountability.

Moreover, as an AI leader, your role is to build trust. Stakeholders must trust that AI will be used transparently, employees must trust that their contributions remain valued in an increasingly automated landscape, and customers must trust that their data is treated with respect and security.

Launch an internal **AI 101 for Leaders** series, short, practical sessions explaining what AI does and doesn't do, using your own data sets as examples. Understanding reduces resistance faster than policy.

Adopt an **Ethics First Charter**. Before any AI build begins, require teams to document data sources, fairness tests, and explainability criteria. Make approval contingent on ethical readiness, not just technical feasibility.

Integrating AI into Strategic Vision

AI is not a tool to be deployed in isolation; it must be woven seamlessly into your organization's strategic vision. This requires a multifaceted approach that addresses infrastructure, goals, and culture.

Building the right infrastructure begins with ensuring data readiness. High-quality, well-structured data is the lifeblood of AI systems. As a leader, you must prioritize investments in data collection, cleaning, and governance to create a foundation for successful AI initiatives.

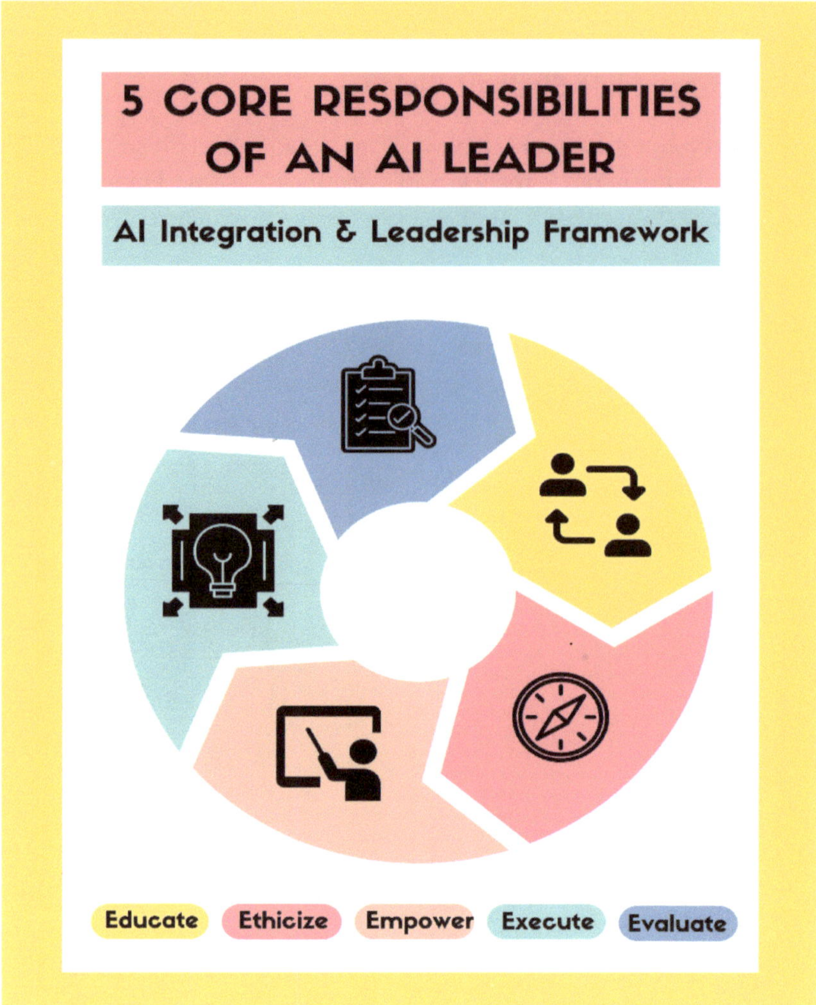

Equally important is aligning AI with clearly defined business objectives. Rather than pursuing AI for its novelty, focus on its ability to address specific challenges or unlock new opportunities. Whether improving customer retention, streamlining supply chains, or enhancing product design, every AI initiative should tie directly to measurable outcomes.

Cultivating a culture that embraces AI is essential. Encourage teams to view AI as an enabler of creativity and innovation rather than a threat. Foster an environment where experimentation is celebrated, and failures are treated as learning opportunities. The more your organization views AI as a collaborator rather than a competitor, the more it will thrive in an AI driven world.

Integrate AI objectives into your annual OKR cycle. Each business unit should own one AI-enabled outcome tied to revenue, efficiency, or experience. This normalizes AI as strategy, not experiment.

Build a **Data Readiness Playbook**. Define data ownership, quality standards, lineage mapping, and refresh cycles. A solid data layer is your AI fuel tank, monitor it like an asset, not an afterthought.

Assembling a High Performing AI Team

The successful integration of AI requires more than technical expertise; it demands a diverse team that blends skills in data science, business strategy, and creative problem solving. Your role as a leader is to assemble and empower this team, ensuring that its members can collaborate effectively and align their efforts with the organization's goals.

Recruiting talent with a strong foundation in machine learning, natural language processing, and data analytics is essential. However, technical skills alone are not enough. Seek individuals with strategic thinking, adaptability, and the ability to translate complex AI insights into actionable business strategies.

Supporting your team's growth is equally critical. AI is a rapidly evolving field, and continuous learning is necessary to stay ahead. Provide access to cutting edge tools, encourage participation in industry conferences, and foster partnerships with academic and research institutions. When your team feels supported and engaged, they are more likely to drive meaningful innovation.

Design an **AI Guild** model, small cross-functional squads pairing data scientists with business and operations experts. Rotate members quarterly to cross-pollinate learning and prevent knowledge silos.

Budget 5–10 percent of your AI project spend for continuous learning. Mandate quarterly skill refresh sessions on emerging frameworks, governance updates, and ethical AI standards.

Overcoming Challenges in AI Implementation

No transformative journey is without obstacles, and AI implementation presents its own set of challenges. From integrating AI with legacy systems to addressing resistance within the organization, these hurdles require proactive and adaptive leadership.

One of the most significant challenges is cultural resistance. Employees may fear that AI will replace their roles or undervalue their contributions. To address this, communicate openly about how AI complements human creativity and expertise. Provide training that empowers employees to leverage AI in their roles, enhancing their capabilities rather than diminishing their importance.

Another challenge lies in navigating the regulatory landscape. As governments worldwide establish guidelines for AI use, organizations must remain compliant without stifling innovation. Stay informed about evolving regulations and collaborate with legal and compliance teams to ensure that your AI initiatives align with both local and global standards.

Technical challenges, such as integrating AI into existing infrastructure, demand a systematic approach. Begin with pilot projects that test AI's capabilities on a smaller scale, gather insights and address issues before scaling up. This iterative process minimizes risk and builds confidence in AI's potential.

Apply a **Pilot-Prove-Scale** rhythm. Launch limited pilots (90 days), prove value with clear metrics, then scale 2× with lessons learned. Avoid leapfrogging to enterprise rollout before cultural readiness.

Use change-enablement playbooks from sectors like public administration and finance where compliance and risk aversion are high; they offer templates for balancing innovation with accountability.

Leading with Ethics and Responsibility

The immense power of AI comes with equally significant responsibilities. Ethical AI leadership requires a commitment to transparency, fairness, and accountability.

Ensure that AI models are explainable, so stakeholders can understand how decisions are made. Address biases in algorithms by implementing rigorous testing and monitoring processes. Safeguard user data with robust security measures and a commitment to privacy.

Encourage open dialogue about the societal implications of AI. Involve diverse voices in discussions about its impact, and remain accountable for the consequences of your decisions. Prioritizing ethics, you not only build trust but also position your organization as a leader in responsible innovation.

Create an **AI Accountability Canvas.** For every initiative, document purpose, data lineage, potential bias sources, audit plan, and human-override points. Review quarterly with risk and HR leaders.

Host Ethics Labs, interactive sessions where cross-functional teams debate real AI dilemmas (e.g., surveillance vs. safety). These conversations humanize governance and keep empathy central.

Measuring the Impact of AI Leadership

The success of AI leadership lies in its outcomes. Effective leaders establish benchmarks for evaluating AI initiatives, focusing on both quantitative and qualitative metrics. Beyond financial performance, consider metrics such as employee engagement, customer satisfaction, and the ethical alignment of AI projects.

Regularly assess the progress of your AI initiatives and remain open to adaptation. Celebrate achievements, learn from setbacks, and maintain a long-term perspective. Successful AI leadership is not static; it evolves with the technology and the needs of your organization.

Deploy an **AI Performance Dashboard**. Combine value metrics (ROI, time saved) with trust metrics (explainability, fairness score, complaint ratio). Share snapshots enterprise-wide to normalize transparency.

After each AI cycle, run a post-implementation review capturing lessons, data drift, and user sentiment. Publish summaries internally, learning builds organizational maturity faster than success alone.

Shaping the Future with AI

The role of an AI leader is not confined to the present. Your decisions today will shape the trajectory of your organization and its impact on the world. Embracing AI with vision and responsibility, you position yourself as a catalyst for progress.

As you navigate the complexities of AI, remember that your leadership extends beyond technology. It is about people, purpose, and progress. Inspire your team to see AI not as a challenge to overcome but as a tool to amplify their potential. Engage stakeholders with a vision that combines innovation with integrity. And lead with confidence that you are shaping a future defined by possibility.

The journey of AI leadership is just beginning. As you harness its power and guide your organization forward, you are not just keeping pace with the future, you are defining it.

Before launching your next AI initiative, ask three framing questions: **Does this solve a real human problem? Can I explain it in one sentence to a non-expert? Would I defend this decision publicly?** If all three are yes, you are leading AI ethically and effectively.

Chapter 21

Leaving a Legacy of Transformative IT Leadership

Assess the core pillars of transformative and enduring leadership.

Transformative leadership in IT is not measured solely by technical accomplishments or immediate successes. The true measure of a leader's impact lies in the legacy they leave behind an enduring influence that shapes organizations, empowers individuals, and fosters a culture of continuous growth and innovation. Leadership that builds a legacy is purposeful and far reaching, creating frameworks that outlast any single tenure.

Your legacy as an IT leader is built on the systems you design, the people you nurture, and the values you embed into the fabric of your organization. It is the culmination of vision, commitment, and action, ensuring your contributions reverberate long after you step away. This chapter serves as a guide to crafting a leadership legacy that inspires, transforms, and endures.

Defining a Legacy of Leadership

A legacy is not about personal accolades or the recognition that comes with leadership roles. It is the collective and lasting impact of the decisions you make, the relationships you cultivate, and the principles you uphold. To define your legacy, start by reflecting on the influence you want to have.

What principles will guide your decisions, and how will they shape the culture of your organization? Consider how your leadership supports long term growth and innovation. A true legacy transcends immediate results, fostering resilience and adaptability that empower others to carry your vision forward.

Your legacy should reflect an alignment between personal values and professional actions. Lead with clarity and purpose, ensuring every initiative, conversation, and decision contributes to the greater mission of your organization and industry.

Legacy Playbook Step 1: Write your **Leadership Legacy One-Liner**. In 15 words or fewer, capture what your impact should make possible for others when you're no longer present. Revisit it monthly to keep decisions aligned.

Legacy Playbook Step 2: Conduct a **Ripple Audit**. Identify 5 decisions you've made in the past year. For each, trace how it influenced others across levels. Use this as a leadership mirror to refine your behaviors.

Embedding Innovation into Culture

Organizations that thrive in the future are those that innovate continuously. As a leader, your role is to embed innovation into the DNA of your organization, ensuring that it remains a core focus long after your tenure. Innovation is not confined to new products or technologies; it extends to how teams work, how challenges are approached, and how opportunities are seized.

Encourage a culture of exploration and experimentation. Foster an environment where employees feel confident taking risks and learning from setbacks. Provide resources and time for creative problem solving, and celebrate successes that emerge from unconventional ideas. When innovation becomes a shared mindset, it creates an enduring impact.

Leadership also involves removing barriers to innovation. Simplify decision making processes, break down silos, and encourage cross functional collaboration. Empower teams to think beyond their immediate responsibilities, cultivating a sense of ownership and possibility.

Field Practice: Schedule a quarterly **Innovation Sprint Week**. Temporarily suspend routine tasks for 2–3 days, empowering teams to explore challenges outside their scope. Require only one output: a practical idea that didn't exist before.

Operationalize innovation through KPIs. Define what **measurable innovation** looks like for each department, not just new tools, but smarter ways of working, better decisions, or faster service delivery.

Empowering the Next Generation

A lasting legacy is defined by the leaders you develop. Mentorship and guidance are powerful tools for fostering growth within your team and ensuring the organization's success beyond your leadership. Invest in identifying and nurturing emerging talent, equipping them with the skills, confidence, and perspective to lead in the future.

Begin with establishing mentorship structures that provide meaningful opportunities for professional development. Share your experiences openly, offering both successes and failures as learning tools. Demonstrating vulnerability and authenticity, you create a space for honest conversations that inspire growth.

Encourage autonomy within your team. Allow future leaders to take ownership of critical projects and make strategic decisions.

Entrusting them with responsibility and providing constructive feedback, you build confidence and competence. The most impactful leaders are those who see their role as cultivating other leaders, not commanding followers.

Start a **Next-Level Leaders Program**. Identify five emerging leaders and assign them high-visibility projects with executive sponsorship. Meet monthly for leadership coaching and reflection.

Use the **Voice Rotation** method. In team forums or planning sessions, periodically step back and ask: "Whose turn is it to lead this conversation?" Create space for your future leaders to grow visibly.

Sustaining Organizational Resilience

Resilience is the hallmark of a thriving organization. A leader's legacy lies in their ability to build structures and cultures that withstand challenges and adapt to change. Focus on developing frameworks that promote sustainability in operations, innovation, and team dynamics.

Systems should be designed with scalability and flexibility in mind. Encourage long-term thinking in decision making, ensuring that immediate benefits do not compromise future opportunities. Whether through adopting advanced technologies, refining processes, or fostering a culture of accountability, resilience must remain central to your leadership approach.

Equally important is cultivating emotional and cultural resilience within your teams. Create an environment where challenges are viewed as opportunities for growth, and provide support during periods of uncertainty. Resilient organizations are those that embrace change with optimism and purpose, guided by leaders who inspire confidence.

Build a **Resilience Operating System** checklist. Document on how your organization handles: disruption, morale dips, leadership transitions, rapid scaling, and service failures. Create playbooks for each.

Introduce reflection rituals like **Challenge Stories** during team meetings, invite one team to briefly share a situation they overcame, what they learned, and what changed. Normalize adaptability as cultural currency.

Extending Impact Beyond the Organization

The influence of transformative IT leadership extends beyond the boundaries of a single organization. Your legacy should include contributions to industry and community, positioning your leadership as a force for broader progress.

Advocate for ethical practices within your sector, particularly in emerging areas like artificial intelligence, data privacy, and sustainability. Use your platform to influence industry standards and promote responsible innovation. Leading with integrity and transparency, you set an example for others to follow.

Community engagement also amplifies your impact. Partner with educational institutions to prepare the next generation of IT professionals. Support initiatives that bridge the gap between education and practical experience, ensuring that diverse perspectives and skills are represented in the field. When you invest in the success of others, your legacy becomes a source of inspiration for countless individuals.

Create a **10× Impact Map**. Identify one way to mentor outside your org, one policy to influence in your sector, and one barrier you'll help dismantle for underrepresented professionals. Track your long-term influence beyond job KPIs.

Launch an open-invitation **Tech Futures Talk** at local universities or community hubs, quarterly conversations on ethics, innovation, and career paths. Show up where future leaders gather.

Aligning Vision with Execution

A powerful legacy requires a clear vision and the ability to translate it into actionable strategies. Vision provides direction and inspiration, while execution ensures that aspirations are realized. Leaders who leave a lasting impact are those who master the art of balancing big picture thinking with practical implementation.

Articulate a compelling vision that resonates with your organization's mission and values. Communicate this vision consistently and passionately, ensuring that every member of your team understands their role in achieving it. Break down the vision into tangible goals, assigning accountability and tracking progress.

Execution requires persistence and adaptability. Regularly assess your strategies to ensure they remain relevant and effective in a rapidly evolving environment. Be open to feedback and willing to adjust to the course when necessary. **When vision and execution align, the results are transformative and enduring**.

Use a **Vision-to-Execution Ladder**. For every aspiration, answer:

> Reflect. Empower. Transition. Every decision you make today shapes the leadership stories of tomorrow.

What outcome would prove this true? What milestone proves we're progressing? Who owns it, and what's next by next quarter?

Run quarterly **Vision Reality Checks**. Share progress toward your big vision using evidence, not enthusiasm alone. Show what's working, what's not, and what's evolving.

Embedding Core Values

Values are the foundation of a lasting legacy. They guide decisions, shape culture, and provide a sense of purpose that endures beyond leadership transitions. Identify the principles that matter most to you whether they are transparency, inclusivity, or innovation and ensure they are embedded in every aspect of your organization.

Lead by example, demonstrating these values in your interactions with teams, stakeholders, and customers. Establish policies and practices that reinforce these principles, creating a culture where they are not just ideals but lived experiences. A values driven organization is one that stands the test of time, guided by a sense of purpose that transcends individual leaders.

Create **Values in Action** logbook. Every month, capture three decisions where your core values directly shaped the outcome. Make it a tool for internal storytelling and external trust.

Ask your team to define your values in their own words. If they can't, your values are decorations. If they can, and if their stories show it, you've built something lasting.

LEADERSHIP LEGACY REFLECTION CYCLE

Reflecting on Your Leadership Impact

A legacy is not built overnight; it is the cumulative result of thoughtful decisions, intentional actions, and unwavering commitment. Regular reflection is essential to understanding the impact of your leadership and identifying opportunities for growth.

Evaluate the systems, culture, and people you've influenced. Assess whether your initiatives have created sustainable value and whether your vision aligns with the evolving needs of the organization. Use feedback from peers, employees, and external stakeholders to gain perspective and refine your approach.

Reflection also involves celebrating successes and acknowledging challenges. Recognize the progress made under your leadership, and take responsibility for areas that require improvement. Embracing self-awareness, you ensure that your leadership remains a source of inspiration and progress.

Set aside a quarterly **Leadership Legacy Reflection Hour**. Ask: **What's working beyond me? What's breaking without me? What story will others tell about this era of leadership?**

Use a 360° legacy survey before transition. Ask direct reports, peers, and external partners one simple question: **What lasting effect has my leadership had on you?**

Preparing for the Transition

Leadership transitions are inevitable, and the way you approach them will define the final chapter of your legacy. A seamless handover ensures continuity, stability, and the ongoing success of your vision.

Develop a robust succession plan that identifies and prepares future leaders. Share your insights and strategies openly, providing the tools and guidance necessary for a smooth transition. Express gratitude to your team and stakeholders for their contributions, emphasizing the collaborative nature of the progress achieved.

Build a **Leadership Runbook**. Document everything, from decision-making rhythms and project cadences to relationship dynamics and unwritten cultural truths. Leave not just tools, but clarity.

Record a 3-minute video for your team titled: **What I Hope We Continue After I'm Gone**. Let your voice anchor the transition, not to preserve control, but to preserve meaning.

A Call to Action

Leaving a legacy of transformative IT leadership is both a responsibility and an opportunity. It requires intentionality, resilience, and a steadfast commitment to creating meaningful change. Reflect on the impact you want to leave behind, and take deliberate steps to shape a future defined by innovation, collaboration, and progress.

Your legacy is not confined to your tenure. It lives in the people you inspire, the systems you create, and the culture you cultivate. Lead with purpose, act with integrity, and empower others to carry your vision forward. In doing so, you leave a legacy that endures, shaping the future of IT and the lives it touches.

Part 8

Leading Through Complexity

(Chapters 22–24)

In an increasingly interconnected and fast paced world, IT leaders face challenges that demand adaptability, foresight, and strategic precision. This part explores the complexities of leading during transformative events like mergers, leveraging data for impactful decisions, and managing distributed teams. It offers guidance to navigate intricate scenarios while maintaining clarity and purpose.

Chapter 22: Leadership During Technology Mergers and Acquisitions

Master the complexities of technology mergers and acquisitions. Learn strategies to align systems, bridge cultural gaps, and ensure seamless integration while maintaining operational continuity. This chapter offers actionable insights for IT leaders managing transformative transitions with precision and clarity.

Chapter 23: Leveraging Data Driven Decision Making

Discover how to harness the power of data to drive strategic decisions. Explore tools, methodologies, and real-world applications that empower IT leaders to turn raw data into actionable insights, aligning business goals with measurable outcomes and fostering innovation.

Chapter 24: Managing Remote and Hybrid IT Teams

Lead effectively in today's dynamic workplace. This chapter provides proven techniques to manage remote and hybrid IT teams, emphasizing communication, collaboration, and productivity. Learn to navigate the challenges of distributed teams while fostering inclusivity and innovation.

HOW TO LEAD THROUGH COMPLEXITY
in Modern IT Leadership

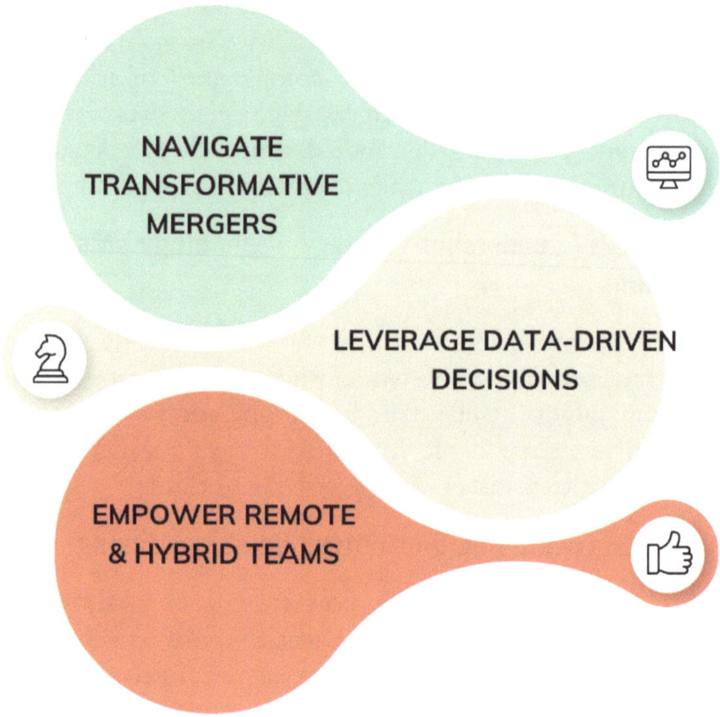

Chapter 22:

Leadership During Technology Mergers and Acquisitions

Leadership during technology mergers and acquisitions is a test of strategic vision, technical expertise, and human centered management. The success of these transitions hinges on a leader's ability to harmonize disparate systems, align organizational objectives, and guide teams through uncertainty while maintaining a steady focus on long term value. Mergers and acquisitions offer transformative potential but require skillful navigation to ensure they deliver on their promise.

This chapter explores how IT leaders can approach mergers and acquisitions with precision and purpose. Through careful planning, effective communication, and a focus on both technical and cultural integration, IT leaders can turn these complex undertakings into opportunities for innovation and growth.

HOW TO LEAD SUCCESSFULLY DURING TECH MERGERS

1. DEFINE CLEAR STRATEGIC OBJECTIVES
2. ALIGN IT INTEGRATION WITH BUSINESS GOALS
3. COMMUNICATE OPENLY TO BUILD TRUST
4. PROTECT DATA AND STRENGTHEN CYBERSECURITY
5. INSPIRE COLLABORATION ACROSS BOTH ORGANIZATIONS

The Strategic Role of IT in Mergers and Acquisitions

The integration of two organizations is often defined by how well their technology ecosystems align. IT leaders are tasked with ensuring operational continuity while positioning the merged entity to achieve its strategic objectives. Technology serves as the backbone of this process, supporting functions from communication and collaboration to data analysis and customer engagement.

IT leaders should begin by understanding the business goals driving the merger or acquisition. Whether the objective is expanding into new markets, reducing costs through synergies, or acquiring cutting edge technology, IT must align its strategies with these aims. This alignment ensures that IT integration efforts are not merely functional but foundational to the success of the merger.

Anticipating challenges is another critical responsibility. Mergers often involve system incompatibility, data integration hurdles, and cybersecurity risks. They are responsible for assessing these challenges early and develop strategies to address them proactively. A merger is more than a technical exercise; it is an opportunity to create a unified, innovative platform that enhances the organization's capabilities.

M&A Playbook Tip: Begin with a **Strategic Systems Impact Map**. For every IT system, identify how it supports a specific business goal of the merger. Color-code systems as 'core to synergy,' 'neutral,' or 'sunset-ready.' Use this map to prioritize efforts.

Strategic Assessment and Planning: Laying the Foundation

The foundation of any successful merger or acquisition lies in thorough assessment and strategic planning. IT leaders are expected to evaluate the technological landscapes of both organizations, identifying strengths, weaknesses, and opportunities for integration.

This process involves taking a detailed inventory of systems, applications, and infrastructure to determine which components are essential and which may become redundant.

Planning must be rooted in clarity and prioritization. Not every aspect of integration can be addressed simultaneously, so they need to focus on the most critical systems first. These may include financial platforms, customer relationship management tools, and operational applications that directly impact business continuity.

Creating a detailed roadmap is essential. This roadmap should outline timelines, milestones, and objectives, while also allowing room for flexibility as unforeseen challenges arise. IT leaders are encouraged to collaborate with stakeholders across departments to ensure that the plan reflects organizational priorities and is feasible within available resources.

Leadership Tool: Run a **60-Day Dual-System Inventory**. Include system owners, license terms, integration readiness, and business dependencies. Tag systems that require early investment vs. late-phase review.

Create a 3-Tier Integration Plan: Tier 1 = must-integrate (within 90 days), Tier 2 = strategic alignment (3–9 months), Tier 3 = optional or sunset (after 9+ months). This reduces overwhelm and aligns stakeholders.

Managing Uncertainty and Fostering Trust

Mergers and acquisitions are often accompanied by uncertainty, particularly among employees who may be concerned about changes to their roles, responsibilities, or job security. For IT teams, these transitions can feel especially daunting, given the complexity and critical nature of their work.

Leadership during this period requires empathy and transparency. They should communicate openly about the integration process, sharing updates, addressing concerns, and providing a clear vision of the future. Regular communication not only alleviates anxiety but also fosters a sense of inclusion and collaboration.

Cultural integration is another significant aspect of M&A leadership. Differences in organizational culture can create friction, particularly when teams are accustomed to different workflows, decision making styles, or communication norms. Leaders need to prioritize building an inclusive culture that respects diversity and promotes a shared sense of purpose.

Create a **Merger Listening Tour**. Meet with key team leads from both orgs. Ask: **What excites you? What worries you? What's sacred in your culture?** Use this feedback to design a merged cultural playbook.

Host bi-weekly **Merger Clarity Briefings** during the first 120 days. Share one decision, one success, one unknown, and one next step. Show visibility and honesty.

Technical Integration: Systems, Data, and Security

Technical integration is one of the most visible and challenging aspects of a merger. IT leaders are responsible for ensuring that systems from both organizations work together seamlessly, supporting operational continuity and enabling strategic growth.

A successful integration begins with evaluating the compatibility of existing systems. They must decide whether to retain one organization's systems, adopt entirely new solutions, or create a hybrid approach. Each decision must be guided by considerations such as scalability, cost efficiency, and alignment with long term objectives.

Data integration is equally critical. Merging data from two organizations require meticulous planning to ensure accuracy, eliminate redundancies, and maintain compliance with privacy regulations. Leaders should develop comprehensive data migration plans that account for cleansing, transformation, and validation processes.

Cybersecurity is a top priority during M&A transitions. The integration process often creates vulnerabilities that can be exploited if not addressed proactively. They are expected to conduct thorough security assessments, implement robust safeguards, and educate employees on best practices to maintain a secure environment.

M&A Tech Tip: Conduct a **Red Team** security simulation between both orgs within the first 30 days. Identify immediate vulnerabilities created by the merger process.

Build a phased **Data Confidence Plan**. Phase 1 = Map and match sources, Phase 2 = Clean and normalize, Phase 3 = Migrate and validate, Phase 4 = Audit and monitor. Assign data stewards from both orgs to co-lead.

Resource Management and Operational Continuity

Balancing the demands of integration with ongoing operations is a key challenge during mergers and acquisitions. Leaders need to ensure that critical day-to-day functions continue uninterrupted while dedicating resources to the integration process.

Resource management begins with identifying and prioritizing key initiatives. They should allocate personnel, budgets, and tools to areas that deliver the greatest value to the organization. Collaboration with vendors and partners can also enhance efficiency, providing access to expertise and resources that facilitate integration.

Maintaining operational continuity requires effective delegation and communication. They are encouraged to empower team members to take ownership of specific tasks, ensuring that no aspect of the integration is overlooked. Regular progress updates and collaborative problem solving help keep the process on track.

Use a **Dual Track** resourcing plan. Track Run-the-Business (RTB) and Change-the-Business (CTB) workstreams separately, with separate metrics and resourcing. This avoids overloading teams.

Empower workstream leads with a **Merger Mission Charter**. For each, define mission, key results, authority boundaries, integration timelines, and cross-functional dependencies.

Post Merger Optimization: Driving Long Term Value

The work of IT leadership does not end with the completion of integration tasks. Post-merge optimization is essential for unlocking the full potential of the merger or acquisition. This phase involves evaluating the performance of integrated systems, identifying opportunities for improvement, and aligning IT initiatives with evolving business goals.

Monitoring key performance indicators is critical during this phase. Metrics such as system uptime, user satisfaction, and cost savings provide valuable insights into the success of the integration. Regular reporting reinforces IT's value to the organization and ensures accountability for achieving strategic objectives.

Continuous improvement is a hallmark of effective leadership. They should foster a culture of innovation that encourages teams to seek out new opportunities for optimization and growth. Remaining adaptable and thinking forward-thinking, leaders can ensure that the benefits of the merger extend well into the future.

Post-Merger Action: Schedule a 100-Day **Integration After-Action Review**. Ask: **What went right? What needs repair? What should scale across the org?** Publish a shared learning doc.

Launch a **Merged Future Labs** team: a cross-org innovation cohort tasked with proposing three optimization initiatives per quarter. Fund rapid pilots and celebrate fast learning.

The Human Element: Inspiring Teams and Building a Legacy

Technology mergers and acquisitions are as much about people as they are about systems. Leaders need to prioritize the human element, inspiring teams to embrace change and contribute to the organization's success. This requires empathetic leadership, clear communication, and a focus on building trust.

Recognizing and celebrating the contributions of team members is essential. M&A transitions are demanding, and acknowledging the hard work of your teams fosters a sense of pride and ownership. Leaders who prioritize their people not only drive successful integrations but also create a lasting legacy of collaboration and achievement.

Celebrate with purpose: Hold a **Merger Unity Milestone** event after major systems go live. Share what each team brought to the table. Capture behind-the-scenes wins and highlight resilience.

Start a **People Before Process** journal: each week, capture one human moment that reminded you leadership is about people, not just platforms. Let it shape how you show up during chaos.

Turning Complexity into Opportunity

Mergers and acquisitions represent a unique challenge for IT leaders, but they also offer unparalleled opportunities for transformation. Approaching these transitions with strategic vision, meticulous planning, and commitment to their teams, Leaders can turn complexity into opportunity.

Success in M&A is defined by the ability to harmonize technical systems, align organizational objectives, and inspire confidence in the future. Those who embrace this challenge with clarity and purpose leave a lasting impact, positioning their organizations for long term growth and innovation.

Chapter 23:

Leveraging Data Driven Decision Making

Data is the lifeblood of modern organizations, fueling innovation, guiding strategy, and driving competitive advantage. For IT leaders, the ability to harness data effectively is essential to making informed decisions that optimize operations, mitigate risks, and uncover new opportunities. However, leveraging data is not simply a technical endeavor; it is a leadership imperative that requires cultivating a culture of curiosity, fostering collaboration, and translating raw information into actionable insights.

This chapter explores how IT leaders can integrate data driven decision making into their leadership approach. Understanding the transformative potential of data, creating an organization that values evidence-based thinking, and addressing the challenges inherent in this journey, you can lead your teams and organization to sustain success.

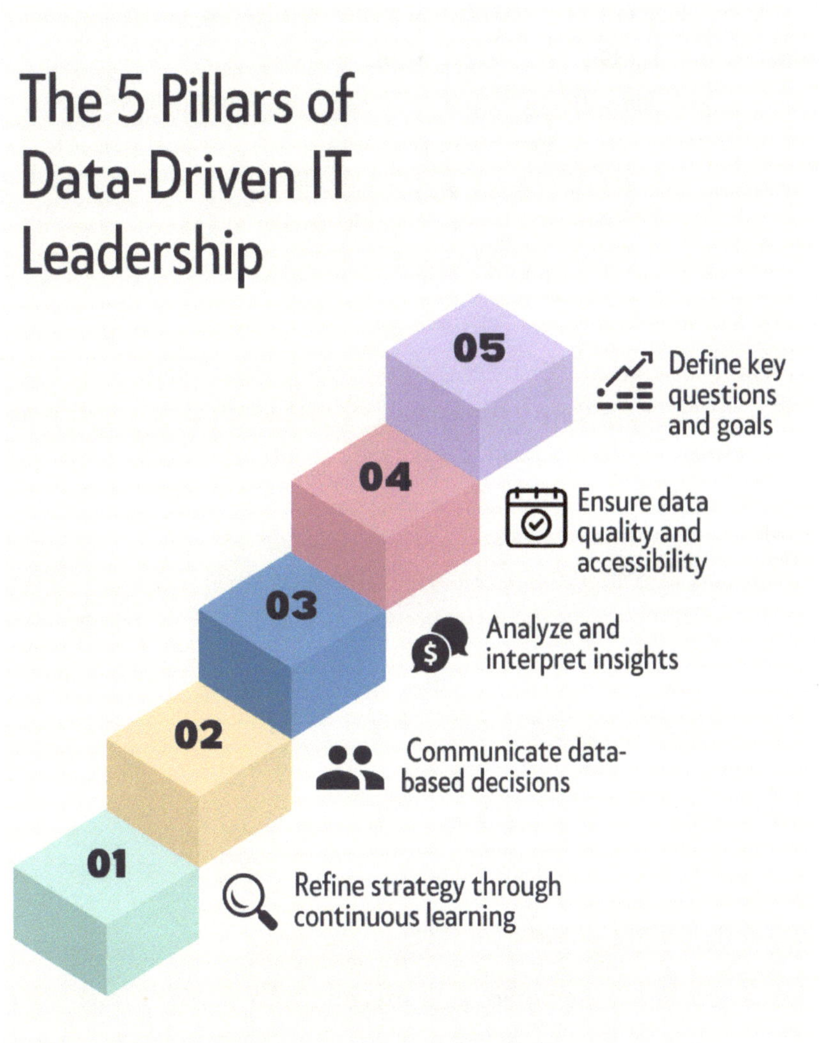

The Transformative Potential of Data in IT Leadership

Data has the unique power to reshape how decisions are made, turning guesswork into precision and intuition into informed action. For IT leaders, data serves as both a tool and a compass, offering clarity in complex environments where decisions often carry significant consequences.

Harnessing data allows IT leaders to move from reactive to proactive leadership. Rather than responding to issues as they arise, data enables the identification of trends and potential risks before they become problems. For example, analyzing system performance metrics may reveal patterns of degradation that could lead to downtime if left unaddressed. With these insights, preventative measures can be implemented, saving time, resources, and reputation.

Beyond operational efficiency, data empowers leaders to shape strategy with greater confidence. Whether optimizing resource allocation, refining project priorities, or identifying customer preferences, data provides the evidence needed to align IT initiatives with organizational goals. It also enhances credibility. Decisions grounded in solid data are easier to justify to stakeholders, fostering trust and reinforcing the strategic value of IT.

Playbook Step: Establish a **Leadership Insights Dashboard** that includes no more than 7 mission-critical KPIs tied directly to your strategic outcomes. Ensure it's refreshed in real-time and reviewed weekly with your leadership team.

Building a Data Centric Culture

For data to fulfill its potential, it must be embedded into the culture of the organization. A data centric culture goes beyond the adoption of analytics tools or technologies; it is a mindset where data is viewed as a critical asset for every decision.

IT leaders are instrumental in fostering this culture. The first step is to ensure that data is accessible and reliable. This requires investments in infrastructure such as data lakes, cloud storage, and advanced analytics platforms that enable seamless collection, management, and analysis of information. However, infrastructure is only part of the equation. Data must also be trusted. Establishing governance frameworks ensures data quality, consistency, and compliance with regulatory standards, making it a reliable foundation for decision making.

Education is equally important in fostering a data centric mindset. Many employees may be unfamiliar or even resistant to using data in their roles. IT leaders should prioritize training programs that demystify analytics and empower teams to harness data effectively. Encouraging collaboration across departments can also break down silos, enabling a more holistic approach to decision making.

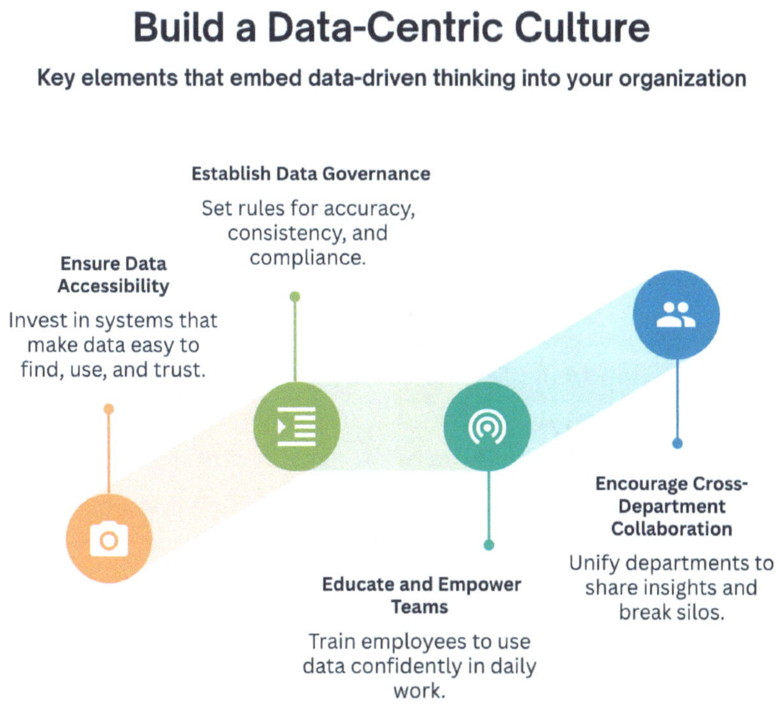

Transparency reinforces the importance of data driven practices. When leaders openly share how data informs their decisions, it demonstrates the tangible benefits of this approach. For example, showcasing how data insights improved customer satisfaction or enhanced operational efficiency can inspire teams to embrace evidence-based thinking.

Leadership Action: Launch a **Data Champions** initiative, one data advocate per team who receives training, resources, and monthly coaching on interpreting, sharing, and applying team-level insights.

Governance Move: Implement a **Data Health Scorecard** system wide. Every major data source should be rated quarterly on freshness, accuracy, completeness, and accessibility.

Turning Data into Actionable Insights

The true value of data lies not in its collection but in its application. IT leaders must be skilled in transforming raw data into actionable insights that inform strategy and execution. This process begins with asking the right questions.

Before analyzing data, define the problems or opportunities that need to be addressed. Clear objectives guide the data collection and analysis process, ensuring relevance and focus. For instance, a leader might seek to understand why a particular system experiences recurring downtime or explore ways to improve user engagement on a digital platform. Well defined questions provide a roadmap for analysis.

With clear objectives in mind, data analysis can begin. Modern tools and techniques, from visualization platforms like Tableau or Power BI to machine learning algorithms, enable leaders to explore patterns, identify correlations, and uncover trends. These insights provide a foundation for decisions that are both informed and impactful.

Communication is a critical step in turning data into action. Insights must be presented in a way that resonates with the audience, whether it's executives, teams, or stakeholders. Visualizations, concise summaries, and narratives that contextualize the data help bridge the gap between analysis and action. Leaders who excel in storytelling can use data to inspire alignment, spark innovation, and drive meaningful change.

Playbook Step: Always follow the **3Cs Insight Model** Clarity (what the data says), Context (why it matters), and Consequence (what action it demands). Train your teams to package insights using this framework.

Tool Tip: Use **Decision Journals** to document which data was used for key decisions, then revisit them in 60 days to validate impact and improve learning loops.

Strategic Data Applications in IT Leadership

Data driven decision making is a versatile tool with applications across every aspect of IT leadership. From enhancing daily operations to shaping long term strategies, data informs choices that maximize efficiency and deliver value.

Operationally, data enables IT leaders to monitor and optimize performance in real time. Metrics related to system uptime, resource utilization, and service response times provide insights that can be used to streamline processes and enhance reliability. Predictive analytics further strengthens operations by identifying potential issues before they occur, enabling preemptive action.

On a strategic level, data supports initiatives such as digital transformation, customer experience enhancement, and innovation. During transformation efforts, data reveals which processes should be prioritized and provides a clear measure of success. For instance, analyzing customer behavior may highlight opportunities to improve engagement through targeted enhancements to digital platforms.

Innovation is another area where data proves invaluable. Examining trends, feedback, and market dynamics, IT leaders can identify unmet needs and design solutions that address them. Data also strengthens risk management efforts, helping organizations assess vulnerabilities, measure the impact of potential threats, and implement effective countermeasures.

Field Practice: Pair each strategic initiative with a **Metrics Readiness Review**. Before launch, ask: What will we measure? How will we collect it? How will we use it to adjust course in real-time?

Advanced Move: Use data not just to report on the past, but to prototype the future, run simulations and scenario planning models before every major strategic decision.

Overcoming Challenges to Data Driven Decision Making

Despite its benefits, implementing data driven decision making presents challenges that IT leaders must navigate. One of the most common obstacles is the presence of data silos, where information is isolated within specific departments or systems. Breaking down these silos requires integration efforts and a culture of open communication that encourages collaboration.

Data quality is another critical concern. Inaccurate, incomplete, or outdated data can lead to flawed analysis and poor decisions. IT leaders must implement robust governance practices, including regular audits and validation processes, to ensure data integrity.

Resistance to change is a cultural barrier that must be addressed through effective leadership. Employees accustomed to traditional decision-making methods may be hesitant to adopt a data driven approach. IT leaders can overcome this resistance by demonstrating the tangible benefits of data driven practices, providing training, and fostering an environment where teams feel supported as they adapt.

Playbook Tip: Run **Data Dust-Offs** cross-team data workshops where participants review underutilized datasets and identify use cases for decision support.

Culture Move: Celebrate **'data wins'** in company comms, showcase a recent decision that delivered measurable value thanks to clear analytics.

Sustaining a Data Driven Future

Data driven decision making is not a one-time initiative but a continuous journey. As technology evolves and organizational priorities shift, IT leaders must remain agile in their approach to data.

Ongoing education is vital to sustaining a data driven culture. IT leaders should invest in professional development opportunities that equip teams with the skills needed to navigate new tools, techniques, and methodologies. Encouraging experimentation fosters a culture of innovation, where teams feel empowered to explore new ways of leveraging data to achieve strategic objectives.

Regularly revisiting data strategies ensures alignment with changing goals and external conditions. Maintaining an iterative approach to data driven decision making, organizations can adapt to emerging opportunities and maintain their competitive edge.

Leadership Habit: Schedule quarterly **Data Strategy Recalibration** sessions. Evaluate which metrics still matter, which insights were most actionable, and which capabilities need to scale.

Sustainability Move: Establish a **Center of Data Excellence,** a lightweight team or hub that curates tools, standards, and success stories for internal use.

Leaving a Legacy Through Data Driven Leadership

Leveraging data effectively is a hallmark of visionary leadership. Embedding data-driven principles into your organization, you create a culture that values evidence, fosters collaboration, and drives meaningful outcomes. The insights derived from data not only optimize performance and mitigate risks but also position organizations for sustained growth and innovation.

As IT leaders navigate the challenges of a digital world, the ability to harness data will define their impact. Those who embrace data driven decision making leave a legacy of informed leadership, agility, and resilience. The journey to data driven excellence is one of continuous learning, adaptation, and exploration a journey every IT leader must undertake to lead with clarity and purpose.

Final Tip: Your legacy is not built by the dashboards you build, it's built by the decisions they empower. Invest in the systems, culture, and people that turn insight into action.

Chapter 24:

Managing Remote and Hybrid IT Teams

The evolving dynamics of work have brought remote and hybrid models to the forefront of organizational structures, reshaping how teams collaborate, communicate, and deliver results. For IT leaders, this shift requires a strategic approach that balances flexibility with accountability, fosters innovation, and sustains team cohesion across distances. Managing remote and hybrid IT teams transcends the logistics of work; it demands leadership that adapts, inspires, and builds resilient systems and cultures to thrive in a decentralized environment.

This chapter explores how IT leaders can address the unique challenges and opportunities of remote and hybrid teams. Leveraging technology, cultivating engagement, and reimagining traditional leadership approaches, you can empower your teams to excel in a work model that defines the future.

Redefining Work: The Transition to Remote and Hybrid Teams

Remote and hybrid work have become more than practical solutions; they represent a fundamental transformation in the relationship between employees and organizations. Advances in digital tools, cloud infrastructure, and communication platforms have enabled teams to collaborate effectively from diverse locations. Employees increasingly value the autonomy and flexibility that remote work provides, while organizations benefit from access to a global talent pool and cost savings associated with reduced physical infrastructure.

For IT leaders, this evolution calls for a redefinition of team management strategies. Remote and hybrid teams introduce complexities such as maintaining communication across time zones, building trust without physical presence, and ensuring seamless workflows in virtual spaces. Addressing these challenges requires a thoughtful balance of technological innovation, empathetic leadership, and organizational agility.

Playbook Tip: Establish a **Work Operating Model Canvas** that clearly defines the expectations, workflows, overlap hours, communication norms, and decision protocols for remote and hybrid collaboration.

Establishing a Foundation of Trust and Clarity

At the heart of effective remote and hybrid team management lies trust. Unlike traditional office environments, where leaders can observe progress directly, remote work demands a shift in focus from processes to outcomes. Leaders must trust their teams to take ownership of their responsibilities while providing the clarity and structure needed to align efforts with organizational goals.

Trust begins with clear communication of expectations. Setting measurable objectives, defining key performance indicators, and outlining project timelines provide team members with a roadmap for success. Frequent check-ins and progress updates create opportunities to reinforce alignment and address any uncertainties. Emphasizing accountability through outcomes rather than micromanagement, leaders foster a culture of mutual respect and autonomy.

Clarity also extends to roles and responsibilities. In hybrid environments where some employees work remotely while others are onsite, it is essential to eliminate ambiguity in task allocation and collaboration protocols. Clearly defined roles reduce redundancies, prevent misunderstandings, and ensure that all team members feel equally valued and engaged.

Leadership Action: Create **Responsibility Maps** for hybrid teams, visually documenting who owns what, shared dependencies, and escalation paths for distributed execution.

Leveraging Technology to Enable Collaboration

Technology serves as the backbone of remote and hybrid teams, enabling communication, collaboration, and productivity across geographies. The choice and implementation of tools directly impact the team's ability to function cohesively. IT leaders must adopt technologies that not only meet technical requirements but also foster inclusivity and efficiency.

Collaborative platforms like Microsoft Teams, Slack, and Zoom facilitate seamless communication, while project management tools such as Asana, Trello, and Jira enable teams to track progress and coordinate efforts effectively. Cloud based solutions ensure that employees can access resources and systems from anywhere, removing physical barriers to productivity.

However, technology must be more than a means to an end; it must be thoughtfully integrated into workflows. Providing training on digital tools ensures that team members are equipped to use them effectively. Periodically reviewing the technology stack allows leaders to identify redundancies, address gaps, and adapt to evolving needs.

Security is another critical consideration. Remote work introduces risks associated with unsecured networks and devices. IT leaders must implement robust security protocols, such as multi factor authentication and encrypted connections, to safeguard organizational data. Regular training in cybersecurity best practices helps team members recognize and mitigate potential threats.

Tool Practice: Implement a **Tech Stack Fit Review** every quarter, survey team usage and satisfaction, identify underutilized tools, and re-align platforms to current workflows.

Security Move: Pair every major tech rollout with a **Secure by Default** checklist. encryption, access control, audit logging, and compliance safeguards as defaults.

Sustaining Team Cohesion and Engagement

One of the most significant challenges of managing remote and hybrid teams is maintaining a sense of connection and shared purpose. The absence of interactions in person can lead to feelings of isolation, reduced morale, and a weakened sense of belonging. To address this, IT leaders must actively cultivate team cohesion and engagement.

Building a positive team culture begins with open and consistent communication. Leaders should prioritize regular team meetings to foster collaboration, share updates, and celebrate achievements. Creating informal virtual spaces such as online coffee chats or team building activities encourages personal connections and strengthens relationships.

Recognition plays a powerful role in sustaining engagement. Acknowledging individual and team contributions reinforces a culture of appreciation and motivates employees to excel. Celebrations of milestones and successes, whether through virtual events or personalized messages, remind team members of their collective impact and value to the organization.

Inclusivity is another cornerstone of engagement. Hybrid models can unintentionally create disparities between remote and onsite employees. IT leaders must ensure that all team members have equal access to opportunities, resources, and information. Equitable practices foster trust and prevent divisions within the team.

Culture Tactic: Run **Inclusion Audits** review how updates, praise, promotions, and decisions are distributed across locations. Make remote workers as visible as onsite contributors.

Connection Habit: Schedule recurring virtual rituals (e.g., "Friday Wins," "First Monday Check-ins," or "Two Truths and a Lie" icebreakers) to humanize distributed work.

Balancing Flexibility with Accountability

Remote and hybrid work models thrive on flexibility, offering employees the autonomy to structure their schedules and environments. However, this flexibility must be balanced with clear accountability to ensure that organizational goals are met.

Establishing performance metrics that focus on results rather than activities allows team members to approach their work in ways that suit their individual strengths and preferences. Regular one-on-one meetings provide opportunities to discuss progress, address challenges, and offer guidance. Emphasizing outcomes, leaders enable employees to maintain their autonomy while contributing effectively to team objectives.

Flexibility also involves accommodating diverse work preferences. While some team members may excel in fully remote environments, others may benefit from periodic in person interactions. Leaders who demonstrate empathy and adaptability create an environment where all team members can thrive, regardless of their work arrangements.

Playbook Step: Use **Results Over Rituals** scorecards, set 3–5 outcome-based goals per role and measure progress over check-in frequency or time spent online.

Adaptability Practice: Offer **Flex Mode Preference** forms, allowing team members to declare when and how they work best (sync vs async, deep focus windows, etc.).

Navigating Challenges and Building Resilience

Managing remote and hybrid teams presents unique challenges, from communication breakdowns to the complexities of coordinating across time zones. Effective leadership requires proactive solutions and a commitment to continuous improvement.

Time zone differences can be mitigated through asynchronous communication and the use of shared calendars to coordinate schedules. Recording meetings and sharing summaries ensure that all team members remain informed, even if they are unable to attend live sessions.

Combating isolation requires a focus on employee wellbeing. Regular check ins allow leaders to understand team members' challenges and provide support. Wellness initiatives, such as encouraging regular breaks or promoting mental health resources, contribute to a positive work environment.

Tactical Tip: Deploy **Timezone Indexing** marks every project stakeholder's timezone next to their name in shared project docs and clarify overlapping windows for live collaboration.

Resilience Booster: Launch optional **Quiet Hours** policies where no meetings are allowed and deep work is protected, reducing burnout from digital fatigue.

Evolving Leadership for a Decentralized Workforce

Leading remote and hybrid teams requires a departure from traditional hierarchical approaches. Leaders must adopt a facilitative style, acting as coaches and mentors rather than authoritative figures. This involves empowering team members to take ownership of their work, fostering a culture of trust, and adapting leadership practices to the needs of a decentralized workforce.

Active listening and empathy are critical skills for navigating the complexities of remote team management. Understanding individual challenges and providing tailored support, leaders can build stronger connections and inspire loyalty. Flexibility in leadership style, whether by adapting communication methods or adjusting management approaches ensures that all team members feel valued and supported.

Leadership Upgrade: Adopt the **Coach-Guide-Remove Barriers** model, support team members with structured feedback, unblock dependencies, and share power through delegation.

Listening Practice: Send a quarterly **Pulse Check** 3-question anonymous form asking: **What's working? What's frustrating? What should we stop/start doing as a team?**

Measuring Success and Embracing Adaptation

The success of remote and hybrid teams can be measured through both quantitative and qualitative metrics. Productivity, project outcomes, and customer satisfaction provide tangible indicators of performance, while employee engagement and retention reflect the health of team culture.

Feedback loops play a crucial role in continuous improvement. Regularly soliciting input from team members allows leaders to identify areas for enhancement, refine processes, and address emerging challenges. A culture of openness and adaptability ensures that remote and hybrid teams remain agile and resilient.

Measurement Move: Track a **Remote Team Health Index** including engagement scores, tool satisfaction, async participation rate, and project velocity trends.

Improvement Habit: Host quarterly **Retro & Rebuild** forums where the team reflects on remote practices, chooses one to retire, and one to evolve.

Shaping the Future of Work

The management of remote and hybrid IT teams is both a challenge and an opportunity for modern leaders. Embracing the unique dynamics of these work models, leveraging technology, and prioritizing empathy and collaboration, IT leaders can unlock the full potential of decentralized teams.

As the workplace continues to evolve, effective leadership in remote and hybrid environments will define organizational success. Fostering a culture of trust, innovation, and resilience, IT leaders can create high performing teams that thrive in a digital first world. In doing so, they not only navigate the complexities of remote work but also shape the future of work itself.

Final Takeaway: Remote work is not just a location shift, it's a leadership shift. Invest in systems, habits, and empathy that scale across distance, diversity, and disruption.

Part 9

Transforming IT Leadership Through Innovation

(Chapters 25–26)

Innovation is the driving force behind IT evolution, and staying ahead requires embracing emerging technologies and frameworks. This part delves into the potential of open-source solutions and the game changing capabilities of quantum computing. It prepares leaders to foster a culture of innovation and align groundbreaking advancements with organizational objectives.

Chapter 25: Leadership in Open-Source Technology Adoption

Unlock the potential of open-source technologies to drive innovation and collaboration. This chapter guides IT leaders in selecting, implementing, and leveraging open-source solutions while addressing integration challenges and fostering a culture of transparency and adaptability.

Chapter 26: Harnessing Quantum Computing in IT Strategy

Prepare for the quantum revolution. Understand how quantum computing can transform industries, solve complex problems, and optimize IT strategies. This chapter explores the fundamentals of quantum technology and its applications, equipping leaders to harness its potential.

Chapter 25:

Leadership in Open-Source Technology Adoption

Open source technology has revolutionized the IT landscape, offering organizations the tools to innovate, adapt, and compete on a global scale. Beyond its technical utility, open source represents a paradigm shift in how technology is developed, shared, and improved. It thrives on collaboration, transparency, and a commitment to collective progress, creating a unique opportunity for IT leaders to drive transformative change.

Adopting open-source technology requires more than technical know-how; it demands strategic leadership, a vision for integration, and a cultural shift that embraces the principles of openness and experimentation. When guided effectively, open-source adoption can become a strategic advantage, accelerating development, reducing costs, and fostering innovation at every level.

Leadership Reminder: Treat open source as a leadership strategy, not just a technical decision. Build an adoption framework that connects business value, community participation, and long-term sustainability.

Leadership Reminder: Treat open source as a leadership strategy, not just a technical decision. Build an adoption framework that connects business value, community participation, and long-term sustainability.

The Rise of Open Source as a Strategic Imperative

Open-source technology has evolved into a critical driver of innovation across industries. It has empowered organizations to adopt solutions that are not only cost effective but also highly customizable and adaptable. From infrastructure management with Kubernetes to advancements in artificial intelligence through TensorFlow, open source tools form the backbone of many modern IT ecosystems.

One of the defining attributes of open source is its collaborative development model. Projects benefit from the contributions of global communities, allowing rapid iteration and enhancement. This distributed model of innovation ensures that open source tools remain at the forefront of technology, often outpacing their proprietary counterparts in flexibility and scalability.

For IT leaders, understanding the potential of open source is key to unlocking its value. This means identifying opportunities where open source can enhance operations, streamline development processes, or enable access to cutting-edge technologies. Open source is more than an alternative to proprietary software; it is a strategic enabler that drives growth, agility, and resilience.

Action Step: Create a **Strategic Open-Source Map**, list existing proprietary systems, identify adjacent open-source equivalents, and evaluate where adoption could drive flexibility or cost savings within 12 months..

Strategic Evaluation of Open Source Tools

The decision to adopt open source technology must align with organizational objectives and technical requirements. Selecting the right tools involves a comprehensive evaluation process that considers compatibility, scalability, and long-term sustainability.

IT leaders must begin by assessing their organization's needs and the specific challenges that open source can address. For example, organizations seeking to modernize their infrastructure might consider tools like Kubernetes for container orchestration or terraform for infrastructure as code. These tools provide the scalability and flexibility necessary to manage complex environments effectively.

Database solutions like PostgreSQL and MongoDB offer organizations robust alternatives to proprietary systems. PostgreSQL is celebrated for its reliability and advanced capabilities, while MongoDB provides unmatched flexibility for handling unstructured data. Both tools empower IT teams to tailor database management to the unique demands of their applications.

Security is another area where open source tools excel. Platforms such as **Wazuh, Zabbix and OpenVAS** provide comprehensive monitoring, intrusion detection, and vulnerability assessment capabilities. These tools allow organizations to maintain a strong security posture while benefiting from the transparency and adaptability inherent to open-source software.

The evaluation process should also account for the maturity of the open-source project. Projects with active communities, comprehensive documentation, and regular updates are better positioned to deliver long term value. IT leaders must assess whether the chosen tools can integrate seamlessly with existing systems and workflows, ensuring that adoption enhances rather than disrupts operations.

Playbook Practice: Run a **5-P Evaluation** before adoption, Purpose, Project Health, People Support, Patch Cadence, and Proven Integration. Approve only tools that pass all five criteria.

Implementation Move: Use a **Governance Readiness Checklist**, confirm license type, patching frequency, dependency chain security, and documentation quality.

Integrating Open Source Technologies into Existing Systems

Integrating open source tools into an organization's technology stack requires meticulous planning and coordination. Unlike proprietary solutions, open source often relies on community driven support, making internal expertise and strategic partnerships essential for successful adoption.

Compatibility is a critical factor in integration. Tools like Prometheus, which provides monitoring and alerting for Kubernetes environments, must be configured to align with existing network, storage, and computing infrastructures. Ensuring compatibility minimizes disruptions and enables organizations to leverage the full capabilities of their open source investments.

Data migration presents another significant challenge. Transitioning from proprietary databases to open source alternatives such as PostgreSQL requires careful planning to preserve data integrity and ensure a seamless transition. This process involves schema mapping, indexing optimization, and comprehensive testing to prevent operational downtime.

To bridge skill gaps, IT leaders should invest in training and certifications. Platforms such as the Linux Foundation, Coursera, and Udemy offer specialized courses on open source tools, equipping teams with the knowledge required to implement and maintain these technologies effectively. Building internal expertise not only ensures smooth integration but also reduces reliance on external support.

Integration Step: Pilot first, launch a 90-day **Open-Source Sandbox** to test interoperability and performance before enterprise-wide rollout.

Capability Tip: Establish internal **Open-Source Guilds** small communities of practice dedicated to continuous learning, documentation sharing, and improvement of open-source tools.

Cultivating a Culture of Collaboration

Open source adoption is as much a cultural shift as it is a technical one. Organizations that succeed in leveraging open source embrace the principles of collaboration, transparency, and experimentation. IT leaders play a crucial role in fostering this culture and encourage teams to explore the possibilities of open source.

Participation in open source communities is one of the most effective ways to embed this culture. Contributing code, reporting bugs, or providing documentation, organizations not only enhance the projects they rely on but also establish themselves as active participants in the global tech ecosystem. This engagement builds credibility and creates opportunities for collaboration with other organizations and developers.

Recognizing and rewarding contributions to open source projects further reinforce this culture. Highlighting success stories and celebrating innovative solutions motivates teams to explore new ideas and challenge conventional approaches. This environment of shared learning and innovation ensures that open source adoption becomes a sustained practice rather than a one-time initiative.

Culture Move: Add **Contribution Hours** dedicate a small percentage of engineering time monthly for community engagement, documentation, or open-source fixes related to tools in use.

Leadership Habit: Feature **Open-Source Spotlights** in internal communications, showcase employee contributions and real project impact to reinforce purpose.

Addressing Challenges and Mitigating Risks

While open-source technology offers numerous benefits, it also presents unique challenges. Licensing compliance, for example, requires careful attention. Open-source licenses, such as GPL, MIT, and Apache 2.0, each come with specific terms and obligations. Leaders are encouraged to ensure that their organization adheres to these requirements to avoid legal complications.

Support and maintenance can also be a concern. Unlike proprietary systems with dedicated help desks, open source relies on community driven support. Building internal expertise and leveraging vendor backed distributions, such as Red Hat Enterprise Linux or MongoDB Atlas, can help address this challenge.

Security is another priority for IT leaders. Although open source projects benefit from community scrutiny, vulnerabilities can still arise. Regular security audits, prompt patching, and the use of tools like Wazuh and OpenVAS ensure that open source environments remain secure. Implementing robust access controls and encryption further enhances protection.

Risk Practice: Form a **License Compliance Registry** maintain a simple catalog listing every open-source component, its license type, and obligations for attribution or redistribution.

Security Step: Implement a **Continuous Vulnerability Watch,** automate scans across all open-source dependencies weekly and track remediation within SLA windows.

Transforming Open Source Adoption into Strategic Advantage

Open source technology has the potential to drive significant strategic advantages for organizations. Tools like Apache Kafka, a distributed streaming platform, enable real time data processing, empowering organizations to make informed decisions quickly. Similarly, TensorFlow and PyTorch democratize artificial intelligence, allowing teams to build advanced models without the constraints of proprietary systems.

Beyond immediate operational benefits, open source fosters innovation by removing barriers to experimentation. Developers can customize tools to meet unique requirements, create prototypes quickly, and iterate based on real world feedback. This agility positions organizations to adapt to changing market conditions and emerging opportunities.

Sustainability is another dimension of open source's strategic value. Tools like OpenDCIM, which provides insights into data center energy usage, help organizations optimize operations for environmental impact. Leveraging these resources, organizations contribute to global sustainability goals while reducing operational costs.

Strategic Play: Develop an **Open-Source Advantage Matrix**, map open-source initiatives to business outcomes like speed-to-market, innovation rate, and sustainability metrics.

Innovation Practice: Encourage **Fork & Prototype** sessions, monthly events where engineers can clone open-source tools, adapt features, and present internal innovation use-cases.

Leaving Legacy of Open Source Leadership

Leadership in open source adoption extends beyond organizational boundaries. Embracing open source principles, IT leaders contribute to a global movement that values transparency, inclusiveness, and collective progress. This leadership leaves a lasting impact not only on the organization but also on the broader tech community.

IT leaders who champion open source adoption create environments where innovation thrives, collaboration flourishes, and barriers to entry are dismantled. Their commitment to advancing open source technology paves the way for future generations of leaders and developers to build on their work.

Open source is not merely a technological choice; it is a strategic philosophy that empowers organizations to lead with vision, act with agility, and inspire through collaboration. Adopting open source with purpose and conviction, IT leaders shape a future defined by shared innovation and enduring impact.

Final Step: Codify your **Open-Source Leadership Charter**, document guiding principles for contribution, transparency, and stewardship. Review it annually as part of your organization's innovation ethics.

Chapter 26:

Harnessing Quantum Computing in IT Strategy

Quantum computing represents a paradigm shift in the field of information technology, promising computational capabilities far beyond those of classical systems. Leveraging the principles of quantum mechanics, quantum computers introduce a new way of processing data that redefines what is computationally possible. At its core, quantum computing relies on qubits, which can exist in multiple states simultaneously, enabling these systems to perform complex calculations at unprecedented speed and scale.

For IT leaders, quantum computing is an emerging force that demands both attention and preparation. While still in its developmental stages, the technology has already demonstrated transformative potential in fields such as optimization, cryptography, and artificial intelligence. Incorporating quantum computing into an IT strategy requires understanding its unique attributes, identifying high impact applications, and preparing for the operational and cultural shifts that accompany its adoption.

Leadership Lens: Treat quantum not as a distant future but as a strategic horizon. Your responsibility is to frame it as a domain worth understanding, investing in, and experimenting with today, not someday.

The Emergence of Quantum Computing as a Strategic Asset

Quantum computing has captured the imagination of technologists, scientists, and business leaders worldwide. Leading companies such as IBM, Google, and Microsoft are at the forefront of quantum innovation, providing platforms that make quantum systems accessible for experimentation and application development. Startups like Rigetti Computing and IonQ are also contributing to a growing ecosystem, advancing quantum technologies through specialized solutions.

What sets quantum computing apart from classical computing is its ability to process and analyze vast combinations of variables simultaneously. This capability arises from quantum phenomena such as superposition, where qubits exist in multiple states, and entanglement, which allows qubits to interact in ways that amplify their computational power. The result is a system capable of solving problems that would be infeasible for even the most powerful classical supercomputers.

The transformative potential of quantum computing spans a variety of industries and use cases. Logistics companies can optimize delivery routes with unparalleled precision, while pharmaceutical researchers can simulate molecular interactions to accelerate drug discovery. As quantum computing continues to mature, its applications will extend to areas that are currently constrained by computational limits, redefining what organizations can achieve through technology.

Action Step: Add a **Quantum Horizon** section in your annual strategy brief, a one-page summary of emerging vendors, pilots underway, or experiments in your industry peers are testing.

Identifying Strategic Applications for Quantum Computing

The strategic value of quantum computing lies in its ability to address complex problems that are computationally intensive or unsolvable using classical systems. IT leaders must focus on aligning quantum capabilities with their organization's goals to identify high priority use cases.

Optimization is one of the most promising applications of quantum computing. Complex problems involving logistics, supply chains, and resource allocation can be addressed more efficiently with quantum algorithms. For example, airlines could optimize flight schedules and fuel consumption to minimize costs while maintaining operational reliability. Similarly, manufacturers could optimize production workflows to maximize efficiency and reduce waste.

In cybersecurity, quantum computing introduces both challenges and solutions. The ability of quantum algorithms to break traditional encryption methods, such as RSA, poses a significant risk to data security. However, quantum key distribution offers a path to unbreakable encryption, providing a new level of security for sensitive communications. Organizations that adopt quantum safe cryptographic methods will be better equipped to safeguard their systems against emerging threats.

Artificial intelligence and machine learning are also fertile grounds for quantum innovation. Quantum systems can process and analyze massive datasets with greater speed, enabling faster model training and more accurate predictions. Financial institutions, for example, could use quantum enhanced algorithms to detect fraudulent transactions in real time or develop sophisticated trading strategies based on market simulations.

Playbook Practice: Host a quarterly **Quantum Use Case Workshop** with stakeholders, use mock cases, discuss operational relevance, and refine top three problems that could benefit from quantum optimization or modeling.

Preparing Organizations for Quantum Computing

Integrating quantum computing into an organization's strategy requires proactive preparation. While widespread adoption may still be years away, early engagement ensures that IT leaders are ready to capitalize on quantum advancements as they become available.

Education is the first step in building quantum readiness. IT leaders and their teams must develop a foundational understanding of quantum principles, including the unique attributes of quantum systems and their potential applications. Resources such as IBM Quantum Experience, Qiskit, and courses offered by academic institutions provide accessible pathways for learning about quantum computing.

Collaboration with research institutions and industry initiatives is another essential step. Partnerships with organizations such as the Quantum Economic Development Consortium enable businesses to access cutting edge research and engage with experts in the field. These collaborations help IT leaders stay informed about technological advancements while identifying opportunities to contribute to the development of quantum applications.

Preparing for quantum computing also involves addressing security concerns. Traditional encryption methods will become obsolete as quantum systems mature, making it critical for organizations to adopt quantum resistant algorithms. Investing in quantum safe cryptographic solutions now will ensure that systems remain secure in the face of future threats.

Readiness Action: Create a **Quantum Readiness Scorecard** include training completion, pilot project pipeline, crypto modernization posture, and partner collaborations. Update annually.

Integrating Quantum Computing into IT Strategy

The integration of quantum computing into IT strategy should be a deliberate and phased process, guided by clearly defined goals and a focus on measurable outcomes. Successful implementation requires balancing exploration with practical application to maximize the value of quantum systems.

The initial phase involves identifying specific problems where quantum computing can deliver significant benefits. IT leaders should prioritize use cases that align with organizational objectives, such as improving operational efficiency, enhancing security, or driving innovation. For example, a pharmaceutical company might explore quantum simulations to accelerate drug development, while a logistics firm could focus on optimizing delivery networks.

Pilot projects provide an opportunity to test quantum capabilities in a controlled environment. Platforms like IBM Quantum and Azure Quantum allow organizations to experiment with quantum algorithms without significant upfront investment in hardware. These pilots enable IT leaders to evaluate the feasibility and impact of quantum solutions, refining their strategies based on real world results.

Hybrid approaches, where quantum systems complement classical infrastructure, offer a practical way to integrate quantum computing into existing workflows. For instance, quantum algorithms can be used to solve specific optimization problems within a broader classical process, leveraging the strengths of both systems.

Pilot Tip: Document each pilot in a **Quantum Impact Log**, capture success indicators, resource usage, algorithm performance, and integration hurdles to shape future rollouts.

Addressing Challenges in Quantum Adoption

The adoption of quantum computing is not without its challenges. The experimental nature of current systems, combined with their high cost and complexity, requires IT leaders to adopt a long-term perspective and manage expectations.

One of the primary challenges is the limited scalability and reliability of current quantum systems. Error rates and qubit limitations mean that practical applications are often constrained. IT leaders must recognize that quantum computing is an evolving technology and plan for gradual integration as capabilities improve.

Cost is another significant factor. Building and maintaining quantum systems requires substantial investment in hardware, talent, and partnerships. Cloud based quantum platforms alleviate some of these costs by providing accessible and scalable solutions. Careful budgeting and alignment with strategic priorities are essential to ensure that investments deliver tangible value.

The scarcity of quantum expertise presents an additional hurdle. Organizations must invest in training and development programs to build internal knowledge while collaborating with quantum experts to bridge gaps in capability. Creating a dedicated quantum team or working with specialized consultants can accelerate the adoption process and ensure that organizations are well positioned to leverage quantum advancements.

Risk Practice: Maintain a **Quantum Technology Watchlist**, track hardware progress, error correction improvements, startup funding rounds, and cryptographic standards in flux.

Investment Guide: Create a staged investment plan with three levels "**explore**," "**experiment**," and "**adopt**" with guardrails for cost, use case criteria, and staffing.

Shaping the Future Through Quantum Leadership

Quantum computing represents a transformative force in the IT landscape, offering unprecedented opportunities to solve complex problems and drive innovation. For IT leaders, embracing quantum computing is not merely about technological adoption, it is about shaping the future of their organizations and industries.

Leadership in quantum computing demands a combination of technical expertise, strategic vision, and adaptability. Understanding the unique attributes of quantum systems, preparing for their challenges, and integrating them into broader IT strategies, leaders can position their organizations to thrive in the quantum era.

The journey to quantum readiness begins with curiosity and a commitment to exploration. As quantum computing continues to evolve, those who lead with determination and foresight will not only adapt to change but actively shape it, leaving a legacy of innovation and progress that defines the next chapter of IT leadership.

Final Action: Declare a **Quantum North Star**, a single sentence mission for how your organization will responsibly explore, test, and apply quantum computing to drive future value.

Part 10

Governance, Ethics, and Resilience in IT

(Chapters 27–30)

Leadership extends beyond technical expertise; it encompasses ethical decision making, governance, and the ability to build resilience in the face of challenges. This part addresses the broader responsibilities of IT leadership, from shaping democratic practices and legal compliance to mastering disaster recovery and negotiation, ensuring a well-rounded, future ready approach.

Chapter 27: The Role of IT Leadership in Shaping Digital Democracies

Explore how IT leaders can promote transparency, inclusivity, and accountability in governance through technology. From open data initiatives to cybersecurity and digital rights, this chapter emphasizes the transformative role of IT in empowering societies.

Chapter 28: The Role of IT in Disaster Recovery Planning

Disaster recovery is critical to resilience. Learn how to prepare for and respond to disruptions with robust strategies, leveraging cutting edge tools and fostering collaboration to safeguard operations, data, and trust.

Chapter 29: The Art of Negotiation for IT Leaders

Negotiation is a cornerstone of leadership. This chapter offers techniques to secure resources, resolve conflicts, and build strong relationships, enabling IT leaders to navigate complex discussions with confidence and achieve mutually beneficial outcomes.

Chapter 30: Legal Challenges and Compliance in IT Leadership

Navigate the evolving legal landscape of IT leadership. From data privacy laws to intellectual property and industry regulations, this chapter equips leaders to address legal challenges proactively, ensuring compliance and protecting organizational integrity.

Chapter 27:

The Role of IT Leadership in Shaping Digital Democracies

Technology has become an integral element of modern governance, creating opportunities to enhance transparency, participation, and accountability. Digital democracies represent a transformative model in which technology is harnessed to empower citizens and modernize governmental processes. From online voting systems to public data sharing platforms, the potential of technology to reshape democracy is profound. However, this transformation requires visionary IT leadership capable of navigating technical complexities, ethical considerations, and societal needs.

The responsibility of IT leaders in this context extends far beyond deploying tools or maintaining systems. It involves fostering inclusivity, ensuring privacy, and safeguarding the democratic ideals that underpin equitable governance. In this chapter, we will explore the role IT leaders play in creating digital democracies that are resilient, ethical, and truly participatory.

Redefining Governance Through Digital Democracies

Digital democracies utilize technology to modernize governance and enhance civic engagement. These systems go beyond digitizing existing processes; they reimagine how citizens interact with government and participate in decision making. Tools such as e voting platforms, real time public feedback mechanisms, and open data portals create a new framework where government becomes more transparent, collaborative, and efficient.

Countries like Estonia offer a clear example of digital democracy in action. Implementing systems like e Residency and online voting, Estonia has enabled its citizens to securely participate in governance from anywhere in the world. Similarly, India's Aadhaar program has established a digital identity system that simplifies access to public services for over a billion people.

These advancements are not without challenges. Issues of data privacy, cybersecurity, and equitable access remain central concerns. IT leaders must design systems that address these challenges while enhancing citizen trust and engagement.

Playbook Action: Initiate a **Digital Democracy Lab** in your agency or region, tasked with piloting open participation platforms and evaluating citizen engagement outcomes.

Building a Strong Digital Infrastructure

A resilient and inclusive digital infrastructure forms the backbone of any digital democracy. IT leaders play a critical role in developing systems that facilitate seamless communication, secure data exchange, and equitable access to digital resources.

Scalability is a vital consideration. Government systems must be capable of handling large volumes of traffic and data, particularly during critical events like elections or emergencies. Cloud platforms provide the scalability needed for these systems while enabling efficient resource allocation. Cybersecurity measures are equally critical, ensuring that sensitive data remains protected from breaches and unauthorized access. Tools such as intrusion detection systems and real time threat monitoring platforms enhance the security of digital infrastructures.

Equity is another key factor. Bridging the digital divide is essential to ensure that all citizens, regardless of socioeconomic status or geographic location, can access and benefit from digital democracy initiatives. Programs that provide affordable internet access, digital literacy training, and multilingual interfaces can help address this challenge. IT leaders should explore innovative solutions, such as leveraging satellite internet or community Wi Fi projects, to bring connectivity to underserved regions.

Infrastructure Directive: Develop a Digital Equity Heatmap to identify underserved populations, broadband dead zones, and accessibility gaps across all major platforms.

Advancing Transparency Through Open Data

Transparency is a fundamental principle of democratic governance, and open data initiatives represent one of the most effective ways to achieve it. Making public data accessible and usable, governments enable citizens, researchers, and organizations to scrutinize and contribute to decision making processes.

Open data platforms create opportunities for innovation and accountability. Governments can use tools like CKAN to build user friendly portals where datasets are readily available to the public. Examples like the European Union's Open Data Portal demonstrate how transparency fosters trust and engagement. Citizens gain insights into government activities, researchers analyze trends, and organizations leverage data to create value for society.

However, transparency must be balanced with the need to protect privacy. Data governance frameworks are critical to ensure that sensitive information remains secure. IT leaders must adopt stringent policies for data anonymization and access control while adhering to international standards for information security. These measures ensure that open data initiatives promote accountability without compromising individual rights.

Implementation Practice: Launch a Public Data Impact Tracker, measuring civic usage, app development, and transparency outcomes per dataset published.

Safeguarding Privacy and Digital Rights

Digital democracies rely on the trust of their citizens, and privacy is central to building and maintaining that trust. IT leaders are tasked with implementing measures that protect user data and uphold fundamental digital rights, such as freedom of expression and equitable access to information.

Strong encryption protocols and secure access controls form the foundation of privacy in digital systems. IT leaders must also address vulnerabilities proactively by conducting regular security audits and deploying advanced tools to detect and mitigate threats. Balancing transparency with privacy is particularly critical in systems that manage sensitive data, such as voter rolls or national identification databases.

Equitable access to information is another key aspect of digital rights. Technology should be accessible to all citizens, regardless of physical ability or economic status. Universal design principles, such as screen reader compatibility and adaptive interfaces, can ensure inclusiveness. Collaborations with organizations that specialize in accessibility can further enhance these efforts.

Policy Tool: Adopt a Digital Rights Charter within your IT governance model, codifying principles of user consent, data minimization, and algorithmic transparency.

Addressing Ethical Challenges in Digital Governance

The integration of technology into governance introduces ethical challenges that require careful consideration. Issues such as algorithmic bias, misinformation, and surveillance have significant implications for the integrity of digital democracies.

Algorithmic bias occurs when automated systems produce unfair outcomes due to flawed data or design. IT leaders must prioritize fairness by auditing algorithms, diversifying datasets, and adopting ethical AI frameworks. Transparency in how decisions are made by algorithms fosters trust and reduces the risk of unintended consequences.

Misinformation is another pressing issue. IT leaders can work with fact checking organizations and develop tools to detect and address the spread of false information. Machine learning models for natural language processing can identify patterns in disinformation campaigns, enabling timely intervention.

Surveillance technologies, while useful for security, must be implemented with strict oversight to prevent misuse. Policies that define acceptable use, along with transparent reporting mechanisms, help safeguard civil liberties.

Ethics Toolkit: Deploy an Algorithmic Impact Assessment for all AI or automation projects, evaluating fairness, accountability, and transparency before deployment.

Empowering Citizens Through Technology

The success of digital democracy hinges on citizen empowerment. IT leaders can drive this empowerment by creating user-friendly platforms that facilitate civic engagement and by investing in digital literacy initiatives.

Online platforms for public participation enable citizens to voice opinions, vote on proposals, and contribute to policy development. Tools like Pol.is and Decide Madrid provide innovative ways to gather collective input and foster collaboration between citizens and governments. These platforms demonstrate how technology can amplify voices and build consensus.

Digital literacy is essential for meaningful participation. Training programs that teach digital skills, data analysis, and cybersecurity empower citizens to navigate and contribute to the digital ecosystem confidently. Partnerships with educational institutions and non-profits can expand access to these resources, ensuring inclusivity.

Civic Action: Establish a **Civic Tech Corps**, a community-driven volunteer program to help residents learn how to use digital government tools and provide feedback loops.

Ensuring Resilience in Digital Democracies

Resilience is a critical attribute of digital democracies. Systems must be designed to withstand threats such as cyberattacks, misinformation campaigns, and natural disasters. IT leaders play a key role in ensuring that digital governance remains stable and reliable.

Comprehensive cybersecurity strategies, including endpoint protection and real time monitoring, are essential for defending against cyber threats. Incident response plans ensure that organizations can recover quickly from disruptions, maintaining trust and operational continuity.

Redundancy in infrastructure further enhances resilience. Backup systems, failover protocols, and distributed networks provide the stability needed to manage unforeseen challenges.

Resilience Practice: Design and rehearse **Digital Governance Stress Tests,** simulations of major civic or election disruptions to measure system and staff response.

Shaping the Future of Governance

The integration of technology into governance represents a profound opportunity to enhance democratic principles and improve societal outcomes. IT leaders are at the forefront of this transformation, driving initiatives that promote transparency, participation, and equity.

Embracing innovation and addressing ethical challenges, IT leaders can ensure that technology serves as a force for good. Collaboration with stakeholders across the public and private sectors amplifies the impact of these efforts, creating systems that empower citizens and strengthen democratic values.

The future of governance is digital, and its success depends on the vision and leadership of those who design and implement its frameworks. Through thoughtful planning, ethical considerations, and a commitment to inclusivity, IT leaders can shape a new era of democracy, one that is transparent, participatory, and resilient.

Legacy Statement: Lead your team in drafting a **Digital Democracy Manifesto**, a public statement of principles and goals your organization commits to as stewards of civic tech.

Chapter 28:

The Role of IT in Disaster Recovery Planning

Disasters, whether natural, human induced, or technological, pose significant risks to the stability and operational continuity of organizations. These events can disrupt infrastructure, compromise sensitive data, and jeopardize critical services. IT leaders hold the pivotal responsibility of developing and implementing comprehensive disaster recovery strategies to protect assets, ensure resilience, and minimize downtime.

Disaster recovery planning is not merely about crisis management; it reflects an organization's commitment to adaptability, reliability, and long-term stability. IT leaders must craft strategies that anticipate risks, integrate advanced technologies, and adapt to an ever-changing landscape of threats. The goal is not only to recover but to lead through recovery and emerge stronger, more agile, and better prepared for future challenges.

Foundations of Disaster Recovery

At its core, disaster recovery involves restoring IT systems, data, and services to their functional state after an unexpected disruption. As a component of business continuity planning, it emphasizes minimizing operational downtime and safeguarding organizational integrity. The effectiveness of a disaster recovery plan is determined by its ability to address the specific vulnerabilities and operational requirements of an organization.

For example, a global financial institution reliant on real time trading systems must prioritize instant data recovery and minimal downtime to maintain market credibility. In contrast, a healthcare provider may focus on continuous access to patient records and life critical systems to ensure uninterrupted care delivery. The key is to design a plan tailored to the organization's unique operational needs and risk landscape.

A robust disaster recovery framework begins with a cross-functional, organization-wide risk assessment and impact analysis. Identifying potential threats such as cyberattacks, natural disasters, and equipment failures, IT leaders can prioritize the resources and strategies required to mitigate those risks. Understanding the dependencies among systems, applications, and data flows provides a holistic view of potential vulnerabilities and informs the design of an effective recovery plan.

Tactical Step: Maintain a live **Critical Systems Map**, and a prioritized recovery dependency matrix across applications, vendors, and infrastructure layers.

Building a Comprehensive Recovery Strategy

A well-defined recovery strategy serves as the blueprint for navigating crises. It outlines the specific steps to restore systems, recover data, and resume critical operations. Clear recovery objectives provide the structure needed to align organizational priorities and technical capabilities.

Two key metrics guide disaster recovery planning: Recovery Time Objective (RTO) and Recovery Point Objective (RPO). RTO defines the maximum allowable downtime for a system, while RPO determines the maximum permissible data loss measured in time. For instance, an e commerce platform may require an RTO of a few hours and an RPO of nearly zero data loss to preserve customer trust and revenue streams. These metrics serve as benchmarks for selecting appropriate technologies, designing workflows, and allocating resources.

Backup and redundancy form the backbone of a recovery strategy. Regular data backups ensure that critical information can be retrieved after a loss, while redundant systems provide fail safe alternatives during infrastructure outages. Many organizations leverage cloud-based solutions to store data in geographically diverse locations, reducing the risk of catastrophic loss. Platforms like Microsoft Azure Backup and Amazon Web Services (AWS) provide scalable, secure options for organizations seeking flexible disaster recovery solutions.

Testing is a vital element of any recovery strategy. Conducting regular disaster recovery drills helps validate the plan's effectiveness, uncover vulnerabilities, and familiarize teams with their roles during an emergency. Simulations that replicate real world scenarios provide valuable insights into potential gaps and ensure that the organization is prepared to execute the plan seamlessly.

Practice Playbook: Run at least two major simulation drills annually, including one full cross-departmental failover scenario and one cyber-focused threat recovery simulation.

Technology's Role in Modern Disaster Recovery

Advances in technology have revolutionized disaster recovery, enabling faster, more reliable, and more efficient recovery processes. Automation plays a pivotal role in minimizing manual intervention and reducing the risk of human error. For instance, tools like VMware Site Recovery Manager orchestrate automated failover and failback processes, allowing critical systems to be restored within minutes.

Cloud computing has redefined disaster recovery by offering on demand scalability, accessibility, and cost efficiency. Disaster

Recovery as a Service (DRaaS) solution, provided by companies like Veeam and Zerto, enable organizations to replicate entire environments in the cloud, ensuring rapid recovery without the need for extensive on premises infrastructure. These solutions also provide flexibility to scale resources based on the severity and duration of the disruption.

Artificial intelligence (AI) and machine learning are emerging as transformative forces in disaster recovery. AI powered tools can monitor system performance, predict potential failures, and optimize recovery workflows. For example, predictive analytics can identify patterns that signal an impending hardware failure, allowing IT teams to take preemptive action. Machine learning models also enhance decision making by analyzing historical recovery data to recommend the most effective restoration strategies.

Technology Stack Recommendation: Evaluate AI-based orchestration platforms that integrate backup, monitoring, and DR automation in a unified console for real-time RTO/RPO tracking.

Addressing the Threat of Cyberattacks

Cybersecurity threats represent a significant challenge in disaster recovery planning. Ransomware attacks, data breaches, and other malicious activities can cripple systems and compromise sensitive information. IT leaders must integrate robust security measures into their recovery strategies to address these risks effectively.

Encryption is essential for protecting data, both during transmission and at rest. Endpoint security solutions, such as those offered by CrowdStrike or SentinelOne, help prevent unauthorized access to systems and detect threats in real time. Additionally, multi factor authentication (MFA) adds a critical layer of security, ensuring that access to sensitive systems is tightly controlled.

Incident response plans complement disaster recovery efforts by defining protocols for detecting, containing, and mitigating cyber threats. Security Information and Event Management (SIEM) systems, such as Splunk or QRadar, provide comprehensive visibility into security events, enabling IT teams to respond swiftly to incidents. Regular penetration testing and vulnerability assessments further strengthen the organization's defenses against evolving cyber threats.

Immediate Control: Include isolated **clean vault** backups that are air-gapped and immutable as part of the ransomware contingency sub-plan.

The Human Dimension of Disaster Recovery

While technology is central to disaster recovery, the human element is equally critical. Effective communication, collaboration, and leadership ensure that recovery efforts are coordinated and efficient. IT leaders must cultivate a culture of preparedness and resilience across the organization.

Clear communication protocols are essential during a crisis. Teams must receive accurate and timely information to make informed decisions and execute their responsibilities effectively. Establishing a command center or crisis management team can centralize decision making and streamline coordination.

Training programs are crucial for ensuring that employees understand their roles in the recovery process. Regular workshops, tabletop exercises, and simulation drills familiarize teams with the recovery plan and build confidence in their ability to respond effectively. Empathy and support from leadership also play a significant role in maintaining morale during challenging times.

Crisis Command Move: Designate three crisis leadership roles, Incident Commander, Communications Lead, and Technical Operations Director, for every major recovery team.

Continuous Improvement and Adaptation

Disaster recovery planning is an iterative process that requires continuous refinement. As technology evolves and new threats emerge, IT leaders must regularly update their plans to remain effective. Incorporating lessons learned from past incidents and industry best practices ensures that the organization stays ahead of potential disruptions.

Feedback loops are an integral part of continuous improvement. After each drill or real-world event, conducting a thorough post mortem analysis helps identify strengths, weaknesses, and opportunities for enhancement. Collaboration with industry peers and participation in forums and conferences further expands knowledge and exposes IT leaders to innovative strategies and tools.

Post-Incident Process: Establish a **Disaster Recovery After Action Report** template to be completed within 72 hours of each event or exercise, including measurable insights and scorecard metrics.

The Strategic Value of Disaster Recovery

Disaster recovery is more than a safeguard against disruptions; it is a strategic asset that demonstrates organizational resilience and reliability. Investing in robust recovery plans, leveraging cutting edge technologies, and fostering a culture of preparedness, IT leaders position their organizations to thrive in the face of adversity.

The role of IT in disaster recovery highlights the transformative power of technology and strategic foresight. It underscores the importance of proactive planning, collaboration, and adaptability in building organizations that are resilient, agile, and prepared for the unexpected. Ultimately, disaster recovery is not just about protecting assets, it is about ensuring the continuity of trust, capability, decision flow, and mission-critical outcomes that define organizational success.

Chapter 29: The Art of Negotiation for IT Leaders

Negotiation is an indispensable skill for IT leaders, enabling them to navigate the complexities of modern organizations, secure critical resources, and foster collaboration across diverse teams. Whether engaging with vendors over contracts, aligning internal stakeholders with IT initiatives, or resolving competing priorities, negotiation is as much about understanding people as it is about achieving strategic objectives.

For IT leaders, negotiation transcends the simple act of reaching agreements. It becomes a sophisticated interplay of preparation, emotional intelligence, technical expertise, and strategic thinking. Mastering this art, IT leaders not only drive organizational success but also strengthen relationships and build a foundation for long term collaboration.

Negotiation in the Context of IT Leadership

Negotiation is integral to the daily responsibilities of IT leaders. From advocating for increased budgets to championing digital transformation initiatives, the ability to negotiate effectively is a critical enabler of progress. In a domain where technology underpins virtually every business function, IT leaders are frequently called upon to act as mediators, problem solvers, and strategists.

Consider the adoption of a new enterprise system. Negotiating favorable terms with a vendor is only part of the challenge. An IT leader must also align diverse internal stakeholders each with unique priorities and concerns to ensure successful implementation. Such situations highlight the multifaceted nature of negotiation, requiring leaders to balance technical knowledge, organizational dynamics, and interpersonal finesse.

Negotiation also plays a key role in achieving alignment between IT goals and broader business objectives. The ability to articulate the value of technology investments in terms that resonate with non-technical stakeholders is often the difference between approval and rejection. Leaders who excel in this area demonstrate not only their technical expertise but also their capacity to build consensus and inspire confidence.

Tactical Move: Position IT proposals in terms of enterprise impact, revenue protection, risk mitigation, speed-to-market, not just technology upgrades.

The Foundations of Successful Negotiation

Preparation is the cornerstone of effective negotiation. It equips IT leaders with the knowledge, clarity, and confidence needed to navigate complex discussions. Comprehensive preparation begins with a clear understanding of objectives. **What are the desired outcomes, and what concessions are acceptable?** Clarifying these points establishes a framework for negotiation strategy.

Research is equally critical. For instance, when negotiating with a vendor, IT leaders must assess market rates, evaluate alternative providers, and understand the vendor's competitive positioning. This data driven approach enables informed decision making and strengthens the leader's bargaining position.

Active listening is another vital component of negotiation. Genuinely understanding the needs, priorities, and concerns of the other party, leaders can uncover common ground and build rapport. Listening with empathy not only facilitates more productive discussions but also establishes trust, a foundational element of any successful negotiation.

Prep Framework: Use the PIA model, Priorities, Interests, Alternatives, to structure pre-negotiation briefings with teams.

Vendor Negotiations: Securing Value Beyond Cost

Negotiating with vendors is a frequent and high-streak task for IT leaders. These discussions often encompass critical aspects such as pricing, service quality, and data security, with outcomes that directly impact organizational performance. While cost considerations are important, successful vendor negotiations prioritize value.

Value extends beyond competitive pricing. It includes factors such as system reliability, scalability, and customer support. For example, selecting a cybersecurity vendor might involve evaluating the provider's ability to deliver rapid response times, comprehensive threat intelligence, and ongoing product updates. These qualitative aspects are often as critical as quantitative pricing terms.

Service level agreements (SLAs) are another focus area in vendor negotiations. These documents define performance benchmarks, ensuring accountability and clarity regarding expectations. Negotiating detailed SLAs, IT leaders can safeguard organizational interests and establish a framework for evaluating vendor performance.

Negotiation Tactic: Always trade concessions, if price flexibility is limited, negotiate for extended support, faster deployment, or feature roadmap influence.

Resolving Internal Conflicts Through Negotiation

Internal conflicts, whether between departments or within IT teams, require a nuanced approach to negotiation. These situations often involve competing priorities, resource limitations, or differences in strategic vision. IT leaders play a pivotal role in mediating such conflicts and fostering collaborative solutions.

For instance, an IT team might advocate for a costly infrastructure upgrade, while the finance department prioritizes cost containment. Resolving this tension requires more than technical justification, it demands empathy, transparency, and creative problem solving. Presenting data driven insights, exploring phased implementation plans, or identifying offsetting cost savings, IT leaders can align stakeholders toward a shared objective.

Transparency is vital in internal negotiations. Sharing relevant data and clearly articulating the rationale behind decisions builds trust and reduces resistance. Inclusive decision-making processes further enhance buy in, fostering a sense of ownership among all parties involved.

Conflict Navigation Tip: Shift internal negotiations from **who wins** to **what's possible together**. Use cross-functional workshops to co-design solutions.

Balancing Technical Expertise and Emotional Intelligence

Effective negotiation in IT leadership hinges on the interplay between technical expertise and emotional intelligence. While a deep understanding of technology is essential for evaluating options and anticipating implications, emotional intelligence enables leaders to navigate interpersonal dynamics with skill and sensitivity.

Consider a scenario where an IT leader negotiates customization requests with a software vendor. Resistance from the vendor might stem from resource constraints or competing priorities. Rather than escalating tensions, a leader with strong emotional intelligence would empathize with the vendor's position, explore mutually acceptable alternatives, and maintain a constructive tone throughout the discussion.

Emotional intelligence is equally critical in managing team dynamics. IT leaders who demonstrate empathy and adaptability are better positioned to address concerns, diffuse conflicts, and inspire confidence. These qualities create an environment of trust and collaboration, enhancing the effectiveness of the negotiation process.

EQ Leverage Strategy: Mirror language, validate emotions, and ask **what would make this a success for you?** to reframe tense conversations.

Harnessing the Power of Data in Negotiations

Data is a powerful asset in negotiations. It provides objective grounding that strengthens arguments and builds credibility. IT leaders who leverage data well can shift discussions from opinion to evidence, increasing the likelihood of successful outcomes.

When negotiating for funding, presenting metrics like ROI, total cost of ownership (TCO), and projected downtime savings makes the business case more compelling. In vendor discussions, data on market trends, benchmarks, and usage forecasts can justify more favorable terms.

Data Strategy: Use visual decision matrices, impact heat maps, and "business risk deltas" to simplify complex tradeoffs in negotiation settings.

The Importance of Flexibility and Creativity

While preparation is essential, successful negotiation also requires flexibility and creativity. Leaders must adapt to the dynamics of each discussion and explore innovative solutions that address the interests of all parties involved.

For instance, if a vendor is unwilling to lower prices, an IT leader might negotiate additional benefits, such as extended warranties, enhanced support services, or flexible payment terms. Similarly, when mediating internal conflicts, proposing compromises such as shared responsibilities or pilot programs can bridge differences and foster agreement.

Flexibility also includes recognizing when to walk away. If a deal fails to align with organizational objectives or values, having the confidence to decline ensures that resources are allocated to more strategic opportunities.

Negotiation Rule: **Never make a deal you wouldn't proudly explain to your successor.** That's how leaders avoid reactive agreements.

Building Long Term Relationships Through Negotiation

Negotiation is not just about short-term gains. It's about building durable relationships that compound over time. Trust, transparency, and mutual respect form the foundation of strong vendors and stakeholder alliances.

IT leaders who negotiate with integrity and pursue win-win outcomes earn reputations as reliable partners. These relationships lead to better service, early access to innovation, and future collaboration opportunities. Maintaining open communication and honoring commitments cements long-term value.

Leadership Mantra: **Negotiate once. Collaborate forever**. Treat vendors and departments like strategic allies, not opponents.

Elevating IT Leadership Through Negotiation

Negotiation is a defining skill for IT leaders, empowering them to drive strategic initiatives, resolve conflicts, and build alliances. Combining technical expertise with emotional intelligence, leveraging data, and fostering collaboration, leaders can navigate even the most complex negotiations with confidence and clarity.

The art of negotiation transcends transactional outcomes. It is a strategic tool for shaping relationships, aligning objectives, and creating enduring values. IT leaders who master this art elevate not only their own effectiveness but also the culture of cooperation and innovation within their organizations.

Chapter 30:

Legal Challenges and Compliance in IT Leadership

The role of IT leadership extends beyond managing technology and aligning it with business strategies, it also demands a deep understanding of the complex legal and regulatory environment in which organizations operate. Legal compliance has become a pivotal responsibility, as technology's centrality in business processes brings heightened scrutiny. From managing data privacy to navigating sector specific mandates, IT leaders are at the forefront of safeguarding organizational integrity while driving innovation and growth.

Compliance is not merely a defensive strategy. It is a proactive commitment that reflects an organization's dedication to ethical governance, accountability, and trustworthiness. This chapter explores the multifaceted legal challenges IT leaders face, offering insights into how they can establish a resilient, legally compliant, and ethically sound digital environment.

Understanding the Legal Landscape for IT Leaders

Legal and regulatory frameworks governing IT operations span a wide array of domains, including data privacy, cybersecurity, intellectual property, and industry specific requirements. These frameworks, which vary across jurisdictions, require IT leaders to maintain a nuanced understanding of obligations to ensure compliance and mitigate risks.

The General Data Protection Regulation (GDPR) in Europe exemplifies the rigorous standards imposed on organizations. It mandates transparency in data collection, explicit consent from users, and prompt breach notifications. Penalties for non-compliance are severe, reaching up to 4% of annual global revenue. Similarly, the California Consumer Privacy Act (CCPA) emphasizes consumer rights, granting individuals control over their personal data.

For organizations operating across borders, the challenges multiply. Differing regulations in various regions demand a cohesive yet adaptable approach. For example, multinational corporations must navigate GDPR's stringent requirements alongside the data access provisions of the U.S. Cloud Act. IT leaders must craft strategies that harmonize compliance across disparate legal environments, leveraging cross functional expertise and technological solutions.

Action Step: Build a global compliance matrix, mapping each regulatory requirement (GDPR, CCPA, HIPAA, PCI DSS, etc.) to your systems, policies, and responsible owners. Use it as a live, continuously updated tool that informs both audits and strategic planning.

In practice: Conduct quarterly regulatory horizon scans in collaboration with legal counsel. This ensures early awareness of new mandates (e.g., AI Act, data residency laws) before they impact operations.

Data Privacy and Protection as Strategic Imperatives

Data privacy is not just a legal obligation but a foundational pillar of organizational trust. IT leaders are responsible for ensuring that data is handled ethically and securely, minimizing risks associated with breaches, misuse, or unauthorized access. This requires implementing policies that adhere to global data protection standards while addressing the unique needs of the organization.

Key practices include encrypting sensitive information during storage and transmission, restricting access to authorized personnel, and conducting regular system audits to identify vulnerabilities. Tools such as encryption software and automated monitoring solutions enable organizations to maintain compliance while enhancing overall security.

Transparency is equally critical. Organizations must provide clear, comprehensible explanations of their data collection practices and offer users meaningful control over their information. Consent management platforms can simplify this process, foster trust and reducing the likelihood of legal disputes.

Incident response is an integral component of data privacy. IT leaders must establish protocols for detecting, containing, and addressing data breaches. This includes notifying affected parties and regulatory authorities within legally mandated timeframes, as required by GDPR and similar laws. Effective incident response not only demonstrates accountability but also mitigates reputational damage.

Playbook Step: Develop a Data Protection Impact Assessment (DPIA) template tailored for your organization. Require it before every new system implementation or vendor engagement.

Execution Tip: Form a **Privacy by Design Task Force** that partners compliance, engineering, and product teams to integrate privacy into system architecture, not as an afterthought, but as a default setting.

In regulated sectors like healthcare or finance, align privacy frameworks with HIPAA, GLBA, and ISO 27701 to ensure harmonized global compliance.

Addressing Cybersecurity as a Legal Obligation

Cybersecurity is both a technical challenge and a legal requirement. Regulatory frameworks like the NIST Cybersecurity Framework and ISO/IEC 27001 provide guidelines for implementing robust security measures. However, compliance extends beyond technical controls, encompassing governance, reporting, and continuous improvement.

IT leaders are tasked with safeguarding critical infrastructure against a diverse array of threats, from phishing attacks to ransomware. This involves deploying advanced threat detection systems, such as intrusion prevention solutions and endpoint security platforms, to monitor and neutralize risks in real time. Regular vulnerability assessments ensure that defenses remain effective against evolving threats.

When breaches occur, legal accountability is paramount. Incident response plans must include detailed protocols for notifying stakeholders and authorities while addressing potential liabilities. For example, a failure to respond appropriately to a ransomware attack could result in fines, lawsuits, and reputational harm. Proactively engaging with legal and compliance teams ensures that responses align with both regulatory requirements and ethical considerations.

Execution Framework: Implement a **Legal Response Workflow** as part of the incident response plan. Assign roles: Legal Counsel (notification), IT Security (containment), PR (communication), HR (internal coordination). Time-stamp each action for audit trails.

Field-Proven Practice: Use tabletop simulations to test cross-departmental responses, simulate a breach in healthcare (HIPAA), finance (SOX), or manufacturing (OT network intrusion) to evaluate readiness and compliance under pressure.

Intellectual Property: Protecting Innovation and Compliance

The management of intellectual property (IP) is central to IT leadership, particularly for organizations that develop proprietary technologies or software. Ensuring the protection of these assets is essential to maintaining competitive advantage and mitigating legal risks.

IT leaders must collaborate with legal teams to secure patents, trademarks, and copyrights for innovations. This involves clearly documenting the development process, maintaining thorough records, and registering intellectual assets with appropriate authorities.

Open-source software introduces additional complexities. While open-source tools provide cost effective solutions and foster innovation, they come with licensing terms that must be carefully adhered to. Non-compliance with licenses such as the GNU General Public License (GPL) or the Apache License can lead to legal disputes and hinder project timelines.

Clear contractual agreements with vendors and contractors are equally important. For example, contracts should explicitly define ownership rights for software developed during engagements, ensuring that the organization retains control over its intellectual assets.

Actionable Tool: Establish an **IP Register** that inventories proprietary code, patents pending, third-party libraries, and open-source dependencies. Update it quarterly with engineering input.

Implementation Tip: Integrate automated license scanners (e.g., FOSSA, Black Duck) into CI/CD pipelines to detect noncompliant code before deployment.

For innovation-heavy industries like manufacturing or energy, standardize invention disclosure processes to accelerate patent filing and protect trade secrets across geographies.

Navigating Emerging Legal Challenges in IT

The rapid evolution of technology continually introduces new legal challenges for IT leaders. Artificial intelligence (AI) and machine learning are at the forefront of this evolution, raising questions about algorithmic transparency, accountability, and bias. IT leaders must ensure that AI systems are designed and deployed ethically, adhering to emerging regulatory standards for fairness and responsibility.

Blockchain technology presents another set of complexities. While its decentralized architecture offers security and transparency benefits, it complicates issues such as data privacy, jurisdiction, and regulatory oversight. IT leaders must carefully assess the legal implications of blockchain adoption and work with legal experts to address potential risks.

Cloud computing is a third area of concern, particularly regarding data sovereignty. Many jurisdictions mandate that data collected within their borders must remain under local control. IT leaders must evaluate cloud providers' compliance with these requirements and implement strategies to mitigate potential conflicts.

Leadership Action: Create a **Responsible AI Governance Board** to oversee model transparency, bias testing, and data ethics compliance. Ensure it includes cross-sector representation, legal, HR, and engineering.

Execution Model: For blockchain and decentralized technologies, deploy legal risk mapping, document where data resides, who controls nodes, and what laws apply.

In highly regulated fields such as finance and public sector, require vendors require compliance with data localization mandates before any contract approval.

Fostering a Culture of Compliance and Accountability

Effective legal compliance requires more than policies and technology, it demands a culture of accountability and collaboration across the organization. IT leaders play a critical role in embedding compliance into the organizational ethos, ensuring that employees at all levels understand their responsibilities.

Training and education are vital. Regular workshops on data protection, cybersecurity best practices, and regulatory standards equip employees with the knowledge needed to navigate legal obligations. Simultaneously, IT leaders must model compliance through transparent decision making and a commitment to ethical practices.

Collaboration with cross functional teams enhances compliance efforts. Involving legal, finance, and operational stakeholders in the planning and implementation of compliance strategies, IT leaders can ensure alignment with broader organizational objectives.

Implementation Tactic: Embed compliance KPIs into leadership scorecards, track training completion rates, audit readiness, and incident closure times as metrics of cultural adoption.

Playbook Step: Appoint compliance champions within each department to act as first responders for policy questions and to flag risk early.

Field Practice Example: In healthcare and public sector, use anonymous "policy pulse" surveys quarterly to measure awareness and identify gaps in understanding regulatory duties.

Leading Through Legal Complexity

The challenges of legal compliance in IT leadership are both intricate and dynamic. From managing global data privacy obligations to addressing the ethical implications of emerging technologies, IT leaders must navigate an ever-changing landscape with foresight and agility.

Success in this domain requires a proactive approach. Staying informed about regulatory developments, leveraging advanced technologies, and fostering a culture of accountability, IT leaders can protect their organizations from legal risks while driving ethical innovation.

Legal compliance is more than a box to be checked, it is a testament to an organization's integrity and commitment to responsible leadership. IT leaders who embrace this responsibility set the standard for excellence, ensuring that their strategies not only meet legal requirements but also inspire trust, resilience, and long-term success.

Leadership Reminder: Treat compliance as strategy, not burden. Build resilience through documentation, transparency, and consistent collaboration with counsel.

Continuous Practice: Conduct annual **compliance retrospectives** to measure progress, document lessons learned and realign priorities with evolving laws.

In essence: Compliance leadership is the art of making ethics operational. Every system, process, and policy should reflect that principle.

Part 11

Strategic IT Execution (Chapters 31–32)

Success in IT leadership hinges on smart infrastructure decisions and effective technology adoption. This part explores must-have tools, infrastructure planning, and vendor consolidation strategies that drive operational excellence, reduce complexity, and future-proof the enterprise.

Chapter 31: Technologies and Tools Every IT Leader Must Adopt

This chapter highlights critical technologies, automation, AI, cloud infrastructure, security, and collaboration, that modern IT leaders must embrace to innovate, stay agile, and lead effectively in today's fast-paced environment.

Chapter 32: Infrastructure Strategy and Vendor Consolidation

Learn how to centralize ISP networks and consolidate vendor relationships for simpler operations, better performance, and scalable IT infrastructure that aligns with business growth.

Chapter 31:

Technologies and Tools Every IT Leader Must Adopt

In today's enterprise technology landscape, tools aren't simply support systems, they are strategic assets. For IT leaders, the choice of platforms spans far beyond individual features; it involves architecture, governance, integration, scalability, vendor trust, and alignment with business outcomes. This chapter maps the essential categories of technology you must master and adopt: from network monitoring and infrastructure, to cyber policy orchestration, Zero Trust, automation and AI, IT service platforms, asset management, observability, collaboration, and open-source/niche innovation. You'll also explore how emerging trends, such as predictive analytics, automated remediation, and unified endpoint management are transforming operational capabilities. Each section includes leading tool examples, evaluation questions, and leadership-level considerations to help you align your technology decisions with broader strategic goals.

1. Network Monitoring & Infrastructure

Every modern IT leader must ensure visibility and control across complex networks, whether hybrid cloud, edge, on-premises or across multiple geographies. Monitoring platforms like Zabbix (open-source), Tenable Nessus (vulnerability and asset exposure), and SolarWinds Orion NPM (enterprise network performance management) are widely adopted.

Key leadership questions include: Do your monitoring solutions allow for centralized dashboarding or are you relying on a fragmented tool-set? Is every site, branch, cloud region and device-type covered? Can you monitor not only uptime and latency, but also configuration drift, network security posture and service-chain impact? Investing in the right network infrastructure tools equip your organization to anticipate issues not just react.

Adopt a **Monitoring Coverage Map** that explicitly lists every location, device type, and platform layer under observation. Update quarterly as your footprint evolves.

Use an industry-proven tactic: tag alerts not just by severity, but by business impact. This prioritizes triage during outages and ties monitoring outcomes directly to service availability.

In regulated sectors like finance or utilities, align network monitoring with compliance reporting frameworks (e.g., PCI-DSS, NERC CIP) to automate evidence collection.

2. Cybersecurity & ePolicy Orchestration (ePO)

Security is now front and center for every business initiative. Platforms like McAfee ePolicy Orchestrator (ePO) bring endpoint policy, patch management, compliance and reporting into a unified console; while endpoint detection and response (EDR), SIEM and vulnerability management products complete the stack. As a leader, focus on: Which security tools form the core of your stack? How systematically are you applying patching at scale? Do you have a centralized policy orchestration layer that feeds upstream dashboards? Security tools aren't just defensive, they're enablers of trust, operational continuity and business agility.

Create a **Policy Enforcement Tier** that spans all devices and integrates into SIEM platforms. This allows policy changes to reflect immediately in reporting and remediation.

Establish patching SLAs by asset class (e.g., critical infrastructure: 48 hrs, remote endpoints: 5 days) and embed those targets into IT team performance metrics.

In public sector or healthcare, integrate compliance dashboards directly into policy orchestrators, so regulatory gaps trigger automated alerts, not just audit surprises.

3. Zero Trust Architecture

The paradigm of "**trust nothing, verify everything**" is now imperative. A Zero Trust framework demands identity-centric, device-centric and application-centric controls. Tools like Zscaler (secure access service edge), Okta (identity management) or Microsoft Defender for Identity may form the foundation of your Zero Trust strategy. From a leadership standpoint ask: Have you defined zero-trust goals (for example, verified device status, identity context, least privilege)? Do your tools support conditional access, monitoring of lateral movement, and adaptive authentication? Embedding zero-trust tools ensures your infrastructure is resilient in a changing threat environment.

Begin with a **Zero Trust Scope Blueprint**, categorize systems into **'initial'**, **'sensitive'**, and **'critical'** zones. This creates a phased rollout roadmap aligned with business risk.

Deploy policy-as-code tools to enforce real-time identity verification and least privilege policies across cloud and on-prem assets.

Apply Zero Trust principles not only to users, but also to service accounts, APIs, and IoT devices, especially in manufacturing and logistics environments.

4. Automation, AI & IT Service Management (ITSM)

Automation and AI transform how IT operations function and scale. Whether it's the ITSM platform like ServiceNow ITSM or remote monitoring & management suites like Atera, automation platforms reduce manual toil and enable responsiveness. Tools such as CyberFOX AutoElevate streamline privilege-management, supporting operational control and security. As a strategic leader, you should explore: Which automation platforms deliver greatest value in your environment? How are you applying AI for incident response, predictive analytics or ticket triage? Automating the repetitive and leveraging AI for the tactical frees your team to focus on innovation, not just upkeep.

Operationalize an **Automation Value Tracker**, a living dashboard that quantifies hours saved, incidents prevented, and MTTR reduced by automation across workflows.

Integrate AI tools into triage stages first, letting algorithms classify, route, and prioritize issues before a human sees them.

Adopt runbook automation platforms for sectors like manufacturing or energy, where time-to-recovery is measured in seconds, not minutes.

5. Help-Desk & IT Service Management

Your help-desk and ITSM platform embody the operational backbone of your IT organization. Whether you utilize ServiceNow, SolarWinds, Freshservice or another tool, success depends on tracking SLAs, ticket lifecycle, user satisfaction and gain from continuous improvement. Leadership must ask: Which help-desk platform do you rely on? Are your KPIs built into the tool and reported at executive level? How does your ITSM platform connect into your automation, asset, and knowledge management ecosystems? A well-functioning help-desk architecture shows your leadership capability in service delivery, not just technology management.

Conduct **SLA Stress Tests** during peak usage periods to validate support resiliency. Adjust staff models and escalation paths based on real-world load.

Deploy self-service knowledge portals directly into help-desk workflows, targeting 20–30% deflection from tier-1 tickets.

For distributed or remote-first teams, ensure multilingual, timezone-aware support routing is embedded into the ITSM stack.

6. Asset & Endpoint Management

Effective IT leadership includes managing the lifecycle of endpoints, licenses and configurations across the ecosystem. Asset management tools provide inventory, deployment, compliance and usage of metrics. Endpoint management solutions help push policies, updates and security patches. Key questions: What asset-management platform tracks devices, licenses and health status? How do you push updates and configuration at scale across desktops, mobile, IoT and remote devices? When you integrate endpoint data into your reporting, you gain control, security and operational clarity.

Deploy a **Single Source of Asset Truth** by integrating asset, endpoint, and licensing data into your ITSM and audit platforms. Automate deprovisioning and lifecycle tagging.

Use real-time endpoint health scores to prioritize patching, especially in high-risk industries such as finance, where device compliance is tied to transaction integrity.

In education or field services sectors, track endpoint compliance by location and job role to align IT controls with operational context.

7. Observability & Incident Response

Traditional monitoring is no longer enough. Observability platforms enable you to understand not just that something broke, **but why it broke**, by analyzing metrics, logs and traces from distributed systems. Platforms such as Datadog, New Relic, Dynatrace and open-source options like Prometheus + Grafana help usher in this deeper visibility. Analysts emphasize that observability must be deployed across infrastructure, applications and business transactions. Leaders should ask: Does your stack cover full-stack visibility? Do you use observability data to drive business outcomes like reduced MTTR or improved user experience? Are you avoiding tool sprawl by consolidating platforms? Observability leadership means turning data into clarity, speed and confidence.

Establish **incident retrospectives** driven by observability data, not opinions. Automate postmortems using timelines from logs and traces.

Consolidate dashboards into role-based views: operations, security, and business impact, all-consuming observability data from a single source of truth.

Use SLIs (Service Level Indicators) and SLOs (Objectives) to translate observability metrics into executive-ready business KPIs.

8. Collaboration & Documentation

In a hybrid or distributed world, technological leadership requires platforms that empower teams, share knowledge and maintain alignment across business and IT. Platforms like Microsoft Teams, Slack, Notion or Confluence support communication and documentation. As a leader you should evaluate: Which collaboration/documentation tools drive transparency across your organization? How do you integrate knowledge management, automation and help-desk workflows? Designing collaboration architecture, you lead not only technology but culture and alignment.

Standardize **Documentation Playbooks** per team, ensuring handbooks, procedures, and project artifacts are continuously updated and reviewable.

Integrate collaboration tools with incident and change management flows, linking Slack messages or Teams chats directly to tickets and resolution logs.

In fast-moving environments, deploy bots that automate knowledge capture (e.g., summarizing sprint discussions into Confluence pages).

9. Open-Source & Niche Tools

Innovation often shows first in open-source and up-and-coming tools. Platforms like Zabbix (network), KeePass (security), OpenProject (project management), or others may provide cost-effective agility. Leadership questions include: Are there open-source or emerging tools your team relies on? What new categories are you tracking (for example AI automation agents, data observability gadgets, micro-services tooling)? Embracing niche tools strategically you gain flexibility, delight engineers and reduce vendor dependency.

Maintain an **Innovation Sandbox** policy, encouraging trial of niche tools under governed conditions before production endorsement.

Assign tech scouts within engineering teams to continuously evaluate emerging GitHub repositories and trending tool chains, then report findings quarterly to leadership.

In budget-sensitive environments like NGOs or academia, open-source solutions can create enterprise-grade capabilities with minimal spending, when governed well.

10. Selection & Leadership Criteria for Enterprise Tools

Selecting tools must be grounded in leadership criteria, not just feature checklists. Evaluate: scalability, integration capability, vendor roadmap and viability, license model, security and compliance posture, alignment with enterprise strategy. Build a decision-matrix that benchmarks prospective tools against these factors, ensures strategic alignment and avoids technology divergence. A powerful leadership move is to embed tool governance, review, retire, update, into your technology lifecycle so your stack continues to serve your evolving mission, not become a burden.

Embed quarterly **Tool Governance Reviews** into your technology planning cycle. Evaluate adoption, duplication, cost efficiency, and alignment to OKRs.

Develop cross-functional selection committees that include engineering, compliance, and business users, ensuring tools meet diverse operational needs.

Use heatmaps to visually track which tools are fully utilized, underused, or redundant, driving optimization and cost containment.

Finally

Technology selection isn't a checklist, it's a leadership instrument. When you align your tool stack with strategic priorities, embed governance, and build execution disciplines around platforms, your IT organization becomes agile, resilient and business-trusted. As you reassess your stack, remember: tools must support vision, not dictate it. As an IT leader, your role is orchestration, ensuring technology works for you, not the other way around.

Leadership Principle: Use tools to unlock leverage, not labor. Every platform should scale outcomes, not just automate steps.

In this chapter, what you've now built is more than a tech stack, it's a blueprint of execution systems that reflect your strategic DNA.

Chapter 32:

Infrastructure Strategy and Vendor Consolidation

As organizations scale across regions, whether nationally or globally, the complexity of managing IT infrastructure grows exponentially. What begins as a handful of locations can quickly evolve into hundreds of sites, each with its own connectivity setup, service providers, hardware standards, and procurement processes. Without a clear infrastructure strategy, this decentralization leads to inconsistent performance, inflated costs, and limited visibility.

Network outages, billing discrepancies, and fragmented vendor relationships become common pain points. Additionally, IT leaders face mounting pressure to support remote operations, enable real-time visibility, and enforce consistent service-level agreements (SLAs) across all touchpoints. As a result, managing infrastructure isn't just a technical function, it's a strategic imperative.

This chapter provides a comprehensive walkthrough of how to design and implement a cohesive infrastructure strategy, with a focus on vendor consolidation and centralized network planning. These two levers, when executed thoughtfully can drive down costs, enhance operational resilience, and empower IT leaders with the visibility and control they need to scale confidently and efficiently.

1. Why Infrastructure Strategy Matters

Whether your enterprise spans four or forty regions, infrastructure must support your organization's growth without creating fragmentation. A cohesive strategy enables consistent delivery of services, faster troubleshooting, standardized policies, and stronger vendor performance management.

A well-architected infrastructure strategy also strengthens compliance and security, especially in industries like healthcare, finance, or retail, where distributed sites often create gaps in visibility and enforcement.

Actionable Step: Create a **Global Network Blueprint** that includes a tiered list of all locations, categorized by strategic priority, compliance obligations, and SLA criticality. This map guides both vendor selection and infrastructure standards.

Use a standard infrastructure intake form for all new sites, whether internal expansion or acquisition. Include questions on connectivity, compliance requirements, hardware baseline, and current vendors. Automate routing this data into your provisioning and NOC systems.

2. Centralizing Network Infrastructure: Two Viable Models

Let's say you operate a healthcare organization like HealthOne, with 6 regional divisions, each covering several U.S. states. Within each state, you manage dozens of clinic locations. This means hundreds of networks, ISPs, routers, and support contracts, making centralized oversight critical.

Here are two proven models to achieve scalable, reliable infrastructure management.

Option A: Aggregated ISP Management via Network Partners (e.g., Granite Telecommunications)

HealthOne's IT leadership chose to partner with a network aggregator like Granite Telecommunications. This model gives them centralized control without forcing them to rip and replace local ISPs.

Here's how it works:

- Granite serves as a single ISP interface across all sites.
- Local ISPs provide last-mile connections, while Granite manages all contracts under a master SLA.
- A dedicated 24/7 Network Operations Center (NOC) receives and handles all service tickets across sites.
- Invoices from 500+ clinics are consolidated into one monthly bill, reducing financial friction.
- SLAs are consistent across all locations, ensuring 99.99% uptime and <50ms latency, even with different regional ISPs.
- Network dashboards like SolarWinds Orion or Zabbix are integrated, giving IT leaders real-time visibility into performance and outages.
- Onboarding new clinics is simple Granite provisions service, assigns a unique ID, and integrates it into the NOC dashboard.

Playbook Insight: Deploy a **Centralized Ticket Normalization Layer**, a lightweight platform or middleware that ensures tickets from all NOC systems are normalized into a common taxonomy for analytics and reporting.

Implement automatic SLA breach escalation: When a site crosses latency or uptime thresholds, auto-alert legal, vendor management, and local tech ops teams for proactive handling.

Healthcare and public sector environments should include compliance SLA monitoring (e.g., HIPAA network standards or public records uptime obligations) as part of aggregator dashboards.

Results:

- Network troubleshooting time dropped by 60%.
- ISP costs decreased by 18% through centralized volume pricing.
- IT maintains control and visibility without needing to manage dozens of different vendor relationships.

Pros:

- Unified management and monitoring.
- SLA consistency across regions.
- Reduced billing complexity.

Cons:

- May include aggregator markup.
- Slightly less flexibility for local ISP negotiations.

Option A: Aggregated ISP Management via Network Partners

Service Aggregation
Combine ISP services through partners

Central Management
Unified control and billing platform

Partner Integration
APIs and SLAs connect providers

Customer Experience
Single support lane and visibility

HealthOne's IT leadership partnered with a network aggregator like Granite Telecommunications to achieve centralized control without replacing existing local ISPs. This model consolidates complexity into a single point of management while preserving regional flexibility.

Single ISP Interface

Granite manages all contracts under one master SLA across all sites.

Unified NOC Support

Dedicated 24/7 Network Operations Center handles all service tickets and escalations globally.

Consolidated Billing

500+ clinic invoices merged into one monthly bill, eliminating financial fragmentation.

Key Performance Metrics: Network troubleshooting time ↓60% | ISP costs ↓18% via centralized volume pricing | 99.99% uptime SLA consistency across all locations with <50ms latency.

→ Real-Time Visibility

Integrated dashboards (SolarWinds Orion, Zabbix) provide immediate visibility into performance and outages across the entire network.

→ Seamless Onboarding

New clinics provision quickly with automatic dashboard integration and unique identifiers assigned by the aggregator.

→ Maintained Control

IT leadership retains full oversight without managing dozens of separate vendor relationships.

✓ Advantages

- Unified management and monitoring across all sites
- Consistent SLAs and performance standards
- Reduced billing and contract complexity

Considerations

- Aggregator markup on pricing
- Reduced flexibility for local ISP negotiations
- Dependency on single aggregator relationship

Option B: Direct ISP National Accounts

Some IT teams prefer to **engage directly with ISPs** like **AT&T, Comcast Business, Spectrum, or Lumen**. This model works well if you have the internal resources to manage vendor relationships.

How it works:

- Negotiate **master service agreements** with major ISPs.
- Enroll each location as a **sub-account** under a national umbrella.
- Use each ISP's **enterprise portal** to manage billing, ticketing, and provisioning per site.
- Maintain internal documentation and monitoring tools to **standardize alerting and performance tracking** across ISPs.
- Negotiate **volume pricing, install SLAs, and support response times** independently with each provider.

Execution Step: Create a shared internal **ISP Relationship Tracker**, a document or tool that records contract renewal dates, escalation contacts, SLAs, and current site coverage for each ISP.

Assign internal regional vendor owners who manage 1–2 ISPs each, supported by quarterly vendor performance review meetings.

For finance, retail, or logistics industries, align direct ISP SLAs with customer-facing uptime expectations, ensuring telecom strategy maps directly to customer impact windows.

Pros:

- Direct relationships with providers.
- Full control over pricing and provisioning.
- Strong leverage for large-scale rollouts.

Cons:

- Higher operational overhead to manage multiple vendors.
- No centralized support center, teams must coordinate with individual ISPs.
- Inconsistent monitoring and response across providers.

When to use this model:

- Your internal team has telecom procurement experience.
- You operate in a few regions where one or two ISPs dominate.
- You want to maintain close control over telecom relationships and performance.

3. Consolidating Distribution and Procurement Channels

Just as with ISPs, **hardware and software procurement** often gets fragmented across vendors, departments, and geographies. To streamline operations:

- Identify **3–5 key distribution partners** (e.g., **CDW**, **PC Connection**, **ARROW**, **Insight**).
- Consolidate purchases under **national contracts** to benefit from pricing tiers and simplified fulfillment.
- Standardize SKUs for laptops, peripherals, networking gear, and licenses across business units.
- Use **procurement platforms** or APIs to integrate purchasing into ITSM or asset management systems (e.g., **Coupa)**.

Strategy: Designate a **Procurement Standards Board** (cross-functional team) that meets biannually to refresh standard hardware lists, assess vendor performance, and remove underused SKUs.

Track lead times, return rates, and warranty claims by distributor, then tie procurement decisions to operational reliability, not just upfront cost.

Benefits:

- Easier inventory tracking and lifecycle management.
- Reduced procurement cycle time.
- Consistent device and software provisioning across sites.

In manufacturing and healthcare, align device standardization with IT compliance (e.g., medical device certifications, ruggedization for industrial use). This reduces operational gaps

Option B: Direct ISP National Accounts

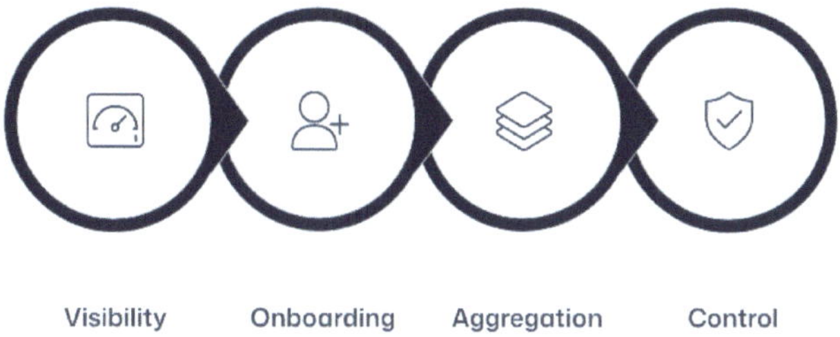

Some healthcare IT teams prefer direct engagement with major ISPs like AT&T, Comcast Business, Spectrum, or Lumen. This approach offers maximum control and negotiating leverage but requires dedicated internal resources to manage multiple vendor relationships effectively.

01

Negotiate Master Agreements

Establish national umbrella contracts with selected ISPs defining pricing tiers, installation SLAs, and support response commitments.

02

Sub-Account Enrollment

Register each clinic location as a sub-account under the national master agreement for streamlined provisioning and billing.

03

Enterprise Portal Management

Utilize each ISP's platform for billing, ticketing, and provisioning while maintaining internal documentation for consistency.

04

Standardized Monitoring

Implement internal tools to normalize alerting, performance tracking, and vendor response across independent ISPs.

✓ Strengths

- Direct provider relationships with full pricing control
- Strong negotiating leverage for large deployments
- No intermediary markup on costs

Trade-Offs

- Higher operational overhead managing multiple vendors
- Decentralized support, coordinate with each ISP separately
- Inconsistent monitoring across providers without central NOC

Best suited for: Teams with telecom procurement expertise operating in limited geographies where 1-2 ISPs dominate, seeking direct control over relationships and performance metrics.

4. Streamlined Infrastructure = Strategic Agility

When IT leaders consolidate infrastructure vendors and streamline network strategy, they unlock:

- **Faster response times** to outages and incidents.
- **Cost savings** through vendor leverage and efficiency.
- **Improved visibility and compliance** across geographies.
- **Scalable onboarding** for new sites, divisions, or acquisitions.

Execution Step: Automate new-site activation by templating provisioning workflows across ISP, firewall, monitoring, and asset tracking layers. Use the template to cut provisioning time by 50–70%.

Embed all infrastructure changes into a quarterly **Infrastructure Governance Cycle** that includes stakeholder review, SLA performance analysis, and change control review across all vendors.

This infrastructure discipline becomes especially valuable during mergers, crises, or rapid growth.

Leadership Principle: Infrastructure is not a sunk cost, it's an agility multiplier. Well-architected infrastructure lets your business pivot faster than competitors, not just operate cheaper.

Consolidating Distribution and Procurement

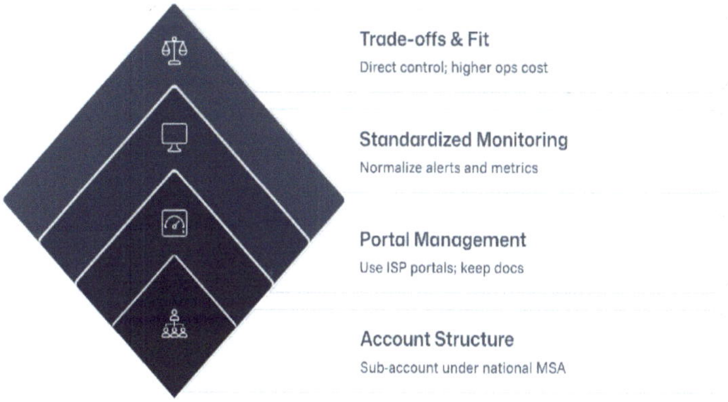

Channels

Beyond ISP management, healthcare IT must address fragmented hardware and software procurement across departments and geographies. Consolidating distribution channels into 3–5 strategic partnerships dramatically reduces complexity and unlocks enterprise pricing.

Key Distribution Partners

Select 3–5 major distributors like CDW, PC Connection, ARROW, and Insight as primary procurement channels.

National Contracts

Consolidate all purchases under national agreements to access volume pricing tiers and simplified fulfillment across 500+ locations.

Standardized SKUs

Define consistent device and software specifications for laptops, peripherals, networking equipment, and licenses enterprise-wide.

1 Streamlined Inventory Management

Centralized procurement enables easier tracking of hardware lifecycle, deployment schedules, and end-of-life planning across all clinics.

2 Reduced Cycle Times

Integration with ITSM platforms accelerates provisioning workflows and reduces time from request to deployment by weeks.

3 Consistent Provisioning Standards

Standardized devices and software configurations ensure security compliance, helpdesk efficiency, and reliable performance across the network.

4 API-Driven Automation

Connect procurement platforms directly to asset management and ITSM systems to eliminate manual order processing and tracking.

Strategic outcome: Combined with aggregated ISP management (Option A), consolidated procurement creates a unified operational model where both network and infrastructure decisions flow through integrated systems, reducing overhead by 30–40% and improving deployment velocity across the healthcare enterprise.

Conclusion

From Visionary to Legend

The journey of an IT leader transcends the boundaries of a career; it is a narrative of impact, transformation, and enduring influence. Every decision, every innovation, and every challenge surmounted contributes to a legacy that defines the future of technology and leadership. Moving from being a visionary to becoming a legend involves more than ambition, it demands authenticity, sustained purpose, and consistent value creation that shapes organizations, industries, and communities.

This journey is a tapestry woven from resilience, purpose, and the ability to inspire others to see beyond limitations. Becoming a legendary IT leader means pursuing the responsibility to lead with integrity, to innovate fearlessly, and to leave a legacy that resonates long after your role is completed.

Embracing a Vision of Impact

Leadership begins with a vision that is both expansive and actionable. A transformative vision reflects the aspirations of a leader who understands the potential of technology to create a better world. It serves as a guiding light, offering clarity in uncertain times and a rallying cry for teams to unite under shared goals.

A compelling vision inspires action and encourages teams to strive for outcomes once thought impossible. Built on principles of innovation, inclusiveness, and sustainability, a strong vision bridges the gap between aspiration and execution. This is the essence of a visionary leader, one who turns abstract ideas into actionable realities and meaningful results.

Courage as the Heartbeat of Leadership

Courage defines the choices that shape your legacy. It is the willingness to take risks, to face adversity head on, and to challenge the status quo for the sake of progress. Courage is not the absence of fear but the ability to act despite it. It enables leaders to advocate for what is right, even when the path forward is fraught with uncertainty.

Leading with courage requires bold decision making and the resolve to navigate complexities without compromising values. It is the determination to transform setbacks into steppingstones and to ensure that every challenge becomes an opportunity for growth. Through courageous leadership, IT leaders pave the way for innovation and resilience, leaving an indelible mark on their organizations.

Elevating Others to Greatness

The most enduring legacies are built not on individual accomplishments but on the success of others. Leadership thrives when teams are empowered to realize their full potential. A legendary leader is not defined by authority but by their ability to foster a culture where people thrive, innovate, and excel.

Empowering others begins with trust, trust in their abilities, ideas, and aspirations. Recognizing contributions, nurturing collaboration, and providing growth opportunities create a culture of mutual respect and shared purpose. Leaders who elevate others build not only high-performing teams but also a legacy of collective success and inspired leadership.

Innovation as a Continuous Pursuit

In a field defined by constant change, innovation is the lifeblood of IT leadership. To leave a lasting legacy, leaders must foster a culture of curiosity, experimentation, and continuous improvement. They must inspire teams to push boundaries, question assumptions, and explore uncharted territories.

Innovation is not a one-time achievement but an ongoing commitment. It is the process of turning ideas into impact, ensuring that organizations remain agile and relevant in an ever-evolving landscape. Championing creativity and embracing calculated risks, leaders position their organizations at the forefront of technological advancement.

Balancing Strategy with Empathy

While technological mastery is essential, the human element of leadership is what transforms strategies into lasting legacies. Leadership is fundamentally about people their aspirations, challenges, and potential. Empathy is the bridge that connects strategic objectives with human experience, ensuring that decisions are both impactful and considerate.

Balancing strategy with empathy involves recognizing the diverse needs of stakeholders and fostering an inclusive environment where every voice is valued. It means leading with humility, listening actively, and prioritizing well-being alongside organizational goals. Leaders who excel in this balance create workplaces where people feel supported, motivated, and connected.

Building an Ethical and Sustainable Legacy

The true measure of a leader is not just what they achieve but how they achieve it. Ethical leadership ensures that decisions are guided by principles of fairness, accountability, and social responsibility. In a world where technology's impact extends beyond the workplace,

IT leaders have a responsibility to advocate for practices that benefit society at large.

Sustainability must be a central tenet of leadership. From championing eco-friendly practices to promoting accessibility and equity, leaders must ensure that their actions contribute positively to the world. A legacy rooted in ethics and sustainability inspires trust and sets a standard for others to follow.

Lifelong Learning and Reflection

Leadership is a journey of growth, not a destination. Lifelong learning is essential for adapting to new challenges, embracing emerging opportunities, and staying ahead of the curve. Leaders who remain curious and open to new perspectives ensure that their strategies remain dynamic and effective.

Reflection is equally important. Taking the time to assess achievements, learning from setbacks, and refining approaches provides a foundation for continuous improvement. Reflecting on the impact of their leadership, IT leaders gain clarity on their values, their vision, and the legacy they wish to build.

Shaping the Future

The journey from visionary to legend is a testament to the transformative power of leadership. It is a journey defined by bold visions, courageous actions, and a commitment to making a difference. Each decision you make shapes not only the trajectory of your organization but also the future of technology and society.

Leadership is not confined to your tenure, it extends through the systems you build, the values you model, and the people you inspire. Your influence carries forward, shaping how others lead, think, and build what comes next.

The future of IT leadership belongs to those who dare to imagine its potential and take decisive action to realize it. Let your legacy be one of courage, compassion, and innovative legacy that inspires others to dream bigger, act bolder, and create a better world.

Appendices

Appendix A: Leadership Frameworks and Tools

Practical guides and frameworks like the Balanced Scorecard, ITIL, and Agile methodologies to provide structure for effective IT leadership practices.

Appendix D: Sustainability Practices in IT

Practical strategies for environmental responsibility, including energy efficient practices, E-waste reduction, and sustainable technology implementations.

Resources for Further Reading

A curated selection of books, research papers, and online courses offering deeper insights into IT leadership, innovation, and global strategies.

Appendix A

Leadership Frameworks and Tools

As an IT leader, the right tools and frameworks turn vision into impact. This appendix highlights critical platforms and industry frameworks that support governance, agility, service delivery, security, collaboration, and growth. Each category includes five curated tools or models for practical applications.

Strategic Alignment & Accountability

- **RACI Matrix** – Clarifies team roles and responsibilities using the Responsible, Accountable, Consulted, and Informed model.
- **COBIT** – IT governance and risk management framework from ISACA.
- **Balanced Scorecard** – Aligns operational activities with organizational vision and strategy.
- **OKR** (Objectives & Key Results) – Goal-setting method linking individual and team efforts to strategic priorities.
- **Lucidchart** – Visual mapping platform for org charts, process flows, and strategy diagrams.

Service Management

- **ITIL** – Comprehensive framework for managing IT services and delivering value.
- **BMC Helix ITSM** – AI-driven service management platform for enterprise environments.

- **ServiceNow** – Workflow-based platform for IT operations and service automation.
- **Cherwell** – Flexible ITSM tool for mid-size and large enterprises.
- Freshservice – Cloud-based, user-friendly IT service desk for agile teams.

Data-Driven Leadership

- **Power BI** – Microsoft's business analytics tool for interactive dashboards and reporting.
- **Tableau** – Advanced data visualization and business intelligence software.
- **Google Data Studio** – Free dashboarding tool for real-time insights.
- **Qlik Sense** – Self-service data analytics with AI and associative exploration.
- **Metabase** – Open-source BI tool for asking questions and visualizing data easily.

Cybersecurity

- **NIST Cybersecurity Framework (CSF)** – U.S. model for identifying and managing cyber risk across five key functions.
- **ISO/IEC 27001** – International standard for information security management systems (ISMS).
- **NIS2 Directive (EU)** – EU law mandating advanced cybersecurity practices for key infrastructure sectors.
- **Cyber Resilience Act (EU)** – Ensures cybersecurity of digital products in the EU market via regulation.
- **CIS Critical Security Controls** – Prioritized U.S.-based guidelines for reducing the risk of widespread cyberattacks.

Project & Change Management

- **Jira** – Agile project management and software development tracking tool.
- **Asana** – Workflow and project coordination platform for goal tracking.
- **Trello** – Visual task management using boards, lists, and cards.
- **ChangeGear** – Change management solution supporting ITIL and digital workflows.
- **Prosci ADKAR** – Change management model for guiding organizational transformation.

Collaboration & Innovation

- **Miro** – Collaborative whiteboard for ideation, strategy, and workshops.
- **MURAL** – Visual thinking tool for remote team collaboration and planning.
- **Excalidraw** – Free, open-source sketching whiteboard for fast idea capture.
- **Figma** – Collaborative UI design and whiteboarding platform.
- **Notion** – Flexible all-in-one workspace combining docs, wikis, and collaboration.

Communication & Engagement

- **Slack** – Real-time messaging platform with channels, file sharing, and integrations.
- **Microsoft Teams** – Communication and video conferencing integrated with Office 365.
- **Zoom** – Widely used video conferencing solution with breakout rooms and webinars.
- **Google Meet** – Video meetings for G Suite users with real-time collaboration.

- **Yammer** – Internal enterprise social network for open communication and feedback.

Knowledge Management & Documentation

- **Confluence** – Team wiki for documenting decisions, processes, and collaboration.
- **Notion** – Centralized workspace for organizing knowledge, tasks, and notes.
- **Obsidian** – Markdown-based note system with backlinking for personal knowledge bases.
- **Coda** – Blends documents, spreadsheets, and apps for unified team workspaces.
- **Slite** – Simple knowledge base for distributed teams to collaborate on documentation.

AI & Automation

- **ChatGPT** – AI-powered assistant for content generation, coding help, and planning.
- **Microsoft Copilot** – Embedded AI within Microsoft 365 to streamline productivity.
- **UiPath** – Robotic Process Automation (RPA) for automating repetitive workflows.
- **Zapier** – Automates workflows between popular web apps with no coding.
- **Make (formerly Integromat)** – Advanced visual automation for connecting complex systems.

Learning & Development

- **Coursera** – Online courses from top universities on tech, leadership, and data.
- **edX** – Learning platform offering professional certifications and degrees.

- **LinkedIn Learning** – Courses covering soft skills, leadership, and IT disciplines.
- **Pluralsight** – Deep technical training for IT professionals and developers.
- **Udemy** – Diverse, affordable online courses on leadership, cloud, and software skills.

Enterprise IT & Software Frameworks

- **TOGAF** – Framework for developing and managing enterprise architecture aligned with business goals.
- **COBIT** – Governance and management framework focused on maximizing IT value and managing risks.
- **NIST Cybersecurity Framework (CSF)** – U.S. cybersecurity standard with global adoption.
- **ISO/IEC 27001** – Global standard for information security and ISMS implementation.
- **Lean IT –** Applies Lean principles to improve IT services, eliminate waste, and increase customer value.
- **Agile Framework** – Iterative project and product management framework promoting adaptability.
- **Scrum** – Agile methodology focused on team-based development through short sprints.
- **CMMI** – Capability maturity model for improving software and process performance.
- **IT4IT** – Reference architecture for managing IT as a business.
- **FEAF** – U.S. Federal Enterprise Architecture Framework for aligning IT with agency missions and strategies.

Appendix B
Sustainability Practices in IT

Sustainable IT is no longer optional; it's a leadership priority. Driving impact starts with energy efficiency. Assess your infrastructure using tools like Google Earth Engine and Schneider Electric's EcoStruxure IT to monitor and optimize energy use in real time. These insights help consolidate workloads and reduce unnecessary hardware.

Cloud platforms like AWS and Azure now offer carbon tracking features, making it easier to choose low-impact configurations. OpenStack remains a flexible open-source option for those seeking control and cost-efficiency. Virtualization further reduces energy demand by allowing multiple systems to run on fewer machines; tools like VMware and Proxmox VE make this accessible and scalable.

E-waste management is critical. Encourage repair and reuse using resources like iFixit, and partner with certified recyclers aligned with standards like R2. On the development side, optimize software with help from the Green Software Foundation, reducing computing loads through efficient coding and lightweight frameworks.

Choose sustainable vendors by reviewing certifications like ENERGY STAR and EPEAT, and use platforms like SupplyShift to assess supply chain responsibility. Internally, platforms such as Culture Amp can help you gather team ideas and foster a culture of environmental responsibility through small, daily habits.

Track your impact with tools like the Carbon Footprint Calculator, turning data into transparency. Leading sustainably in IT means embedding responsible choices at every level from procurement to design to daily operations. These tools and strategies are a starting point for lasting, measurable change.

Resources for Further Reading

IT leadership demands continuous learning. Staying ahead means mastering frameworks, understanding trends, and drawing insights from experts. This curated list of resources offers practical guidance, strategic tools, and fresh perspectives to support your ongoing development.

Books:

The Phoenix Project by Gene Kim, Kevin Behr, and George Spafford – A story-driven introduction to DevOps, teamwork, and transforming IT operations.

Measure What Matters by John Doerr – A practical guide to using OKRs for goal alignment, focus, and measurable results.

Cybersecurity and Cyberwar by P.W. Singer and Allan Friedman – A clear, accessible overview of today's cybersecurity challenges and strategies.

ITIL Foundation: ITIL 4 Edition by AXELOS – The official guide to ITIL, essential for aligning IT services with business goals.

Leadership in IT by Various Authors – A collection of insights on innovation, collaboration, and leading in complex, tech-driven environments.

The Lean Startup by Eric Ries – A blueprint for fostering innovation through experimentation, iteration, and adaptability.

These resources are your companions for the journey ahead offering the knowledge and tools to lead with clarity, resilience, and impact.

Your Action Plan

An impactful IT leader doesn't just absorb knowledge; they take deliberate action to translate insights into results. This action plan is your blueprint for applying the principles, frameworks, and tools shared throughout this book. It's designed to be both practical and adaptable, guiding you step by step toward creating meaningful changes in your organization and career.

Step 1: Define Your Vision

A clear vision is the cornerstone of effective leadership. Start by identifying what you want to accomplish, both for your team and your organization. This involves aligning your IT goals with the larger business objectives. For example, if your company is focused on enhancing customer experience, your IT vision might include implementing a more responsive digital infrastructure or adopting AI driven analytics tools. Use frameworks like OKRs (Objectives and Key Results) to define actionable, measurable goals. Free tools like Google Sheets or Airtable can help you map out your objectives and track progress. A strong vision should inspire confidence, provide direction, and serve as a benchmark for decision-making.

Step 2: Assess Your Current State

Before moving forward, you need to understand where you stand. Conduct a thorough assessment of your IT infrastructure, workflows, and team capabilities. Use free tools like Zabbix for performance monitoring or OpenProject for analyzing task completion and resource allocation. Evaluate critical areas such as cybersecurity, system uptime, service delivery, and team communication. Don't overlook user satisfaction and support ticket trends, which often signal deeper operational issues. Engage your team in this process, as their insights often reveal hidden challenges or opportunities.

Step 3: Prioritize Key Initiatives

With an understanding of your current state, focus on the initiatives that will deliver the most value. Avoid spreading your efforts too thin by tackling too many projects at once. Instead, prioritize is based on factors like business impact, cost, and resource availability. For example, if cybersecurity is a pressing concern, consider implementing tools like Wazuh for intrusion detection and compliance management. Platforms like Trello or Asana (which offer free versions) can help you visually organize and prioritize these initiatives. Use MoSCoW or RICE frameworks to weigh effort versus value. Breaking larger projects into smaller, actionable tasks ensures steady progress and allows you to pivot as needed.

Step 4: Build Your Team

A successful leader recognizes the importance of a strong, well-prepared team. Start by evaluating the skills and expertise of your current team members. Identify gaps and create development plans to address them. Platforms like LinkedIn Learning and Coursera provide free or affordable courses in cybersecurity, cloud computing, project management, and more. Pair junior employees with experienced mentors using tools like MentorcliQ or open-source mentorship platforms. Foster cross-functional training to increase agility and reduce reliance on single points of expertise. Investing in your team's growth builds a resilient workforce ready to take on tomorrow's challenges.

Step 5: Implement Small, Strategic Wins

Early successes can set the tone for larger achievements. Identify areas where you can quickly implement improvements to show progress and build confidence. For instance, improving Mean Time to Resolution (MTTR) with free tools like OTRS can demonstrate immediate operational benefits. Small wins help create momentum, boost morale, and establish credibility with stakeholders. Use these successes to gather feedback, build trust, and refine your approach to more complex projects. Document these wins with before-and-after metrics to showcase impact.

Step 6: Monitor and Measure Progress

Success isn't static, it needs to be measured and adjusted. Identify key performance indicators (KPIs) that align with your goals and monitor them consistently. Use tools like the Green IT Calculator for sustainability metrics or Google Data Studio to build dashboards that track operational and strategic KPIs. Incorporate service desk analytics, health system, and cost optimization metrics to gain a comprehensive view. Review performance regularly with your team to identify trends, improve accountability, and make informed decisions.

Step 7: Communicate and Inspire

Leadership goes beyond managing systems, it's about inspiring people to contribute their best. Share your vision and progress using clear, compelling language. Frame communication around the impact of your initiatives, emphasizing how they support organizational goals. Use storytelling techniques to connect emotionally and intellectually. Whether presenting a dashboard in Tableau or walking through a user success story, effective communication ensures alignment and fosters motivation. Be consistent across channels, use Slack, internal wikis, and leadership meetings to reinforce your message.

Step 8: Continuously Learn and Adapt

In the fast-paced world of IT, standing still means falling behind. Dedicate time for continuous learning to stay ahead of trends. Free platforms like edX, MIT OpenCourseWare, and Google Digital Garage offer relevant content in emerging tech, leadership, and business strategy. Encourage your team to do the same by providing access and time for development. Establish regular "learning sprints" or "tech talks" within your team to share knowledge and lessons learned. Revisit and refine your strategies based on new insights and evolving circumstances.

Step 9: Focus on Sustainability

Sustainability isn't just a trend, it's a responsibility. Evaluate the environmental and long-term operational impact of your IT systems. Use tools like the AWS Sustainability Dashboard or Green IT Calculator to identify areas for improvement. Transition to cloud-native architectures or energy-efficient hardware where possible. Consider lifecycle management for devices and sustainable procurement practices. Aligning your IT roadmap with ESG (Environmental, Social, and Governance) goals strengthens brand reputation and reduces long-term costs.

Step 10: Reflect on Your Impact

At the end of each initiative or strategic cycle, take time to reflect on outcomes and growth. Engage your team in open retrospectives to collect feedback and assess what worked, and what didn't. Evaluate how your actions have contributed to organizational success and your leadership development. Capture key takeaways in a shared knowledge base for future reference. Reflection ensures lessons are carried forward and help shape a vision grounded in experience. Leadership is not only about results, but also about leaving a meaningful, lasting imprint.

Checklist for IT Leadership Excellence

Excellence in IT leadership demands clarity, agility, and execution. This checklist helps you evaluate the critical dimensions of your leadership, ensuring you're aligned with the demands of a fast-evolving digital environment. Use it to identify strengths, spot gaps, and drive impactful progress.

Vision and Strategic Alignment

5. Is your IT vision clearly articulated and aligned with company-wide goals?
6. Are strategic objectives actionable and measurable using tools like OKRs?
7. Do you regularly communicate the vision in a way that energizes stakeholders?
8. Are technology initiatives directly tied to business growth or transformation?
9. Do you revisit your vision periodically to reflect industry and internal changes?

Team Development and Culture

1. Have you assessed individual and team-wide skill levels and growth areas?
2. Are there structured learning paths via platforms like Coursera or LinkedIn Learning?
3. Is there a mentorship or coaching framework in place?

4. Does the team environment support innovation, autonomy, and collaboration?
5. Do you recognize and reward contributions that align with strategic goals?

Governance and Compliance

1. Are you aligned with industry frameworks such as COBIT, ITIL, ISO/IEC 27001, NIST CSF, or TOGAF for enterprise architecture and risk management?
2. Is compliance with regulations like GDPR, HIPAA, SOX, PCI DSS, CCPA, or FedRAMP proactively managed?
3. Do you conduct regular audits, maintain policy documentation, and follow change control procedures?
4. Are risk management protocols integrated into daily operations and decision-making workflows?
5. Is governance designed to enable accountability and transparency without creating operational bottlenecks?

Operational Performance and Metrics

1. Do you track metrics like MTTR, uptime, ticket volume, and NPS?
2. Are these KPIs reviewed consistently with your team and stakeholders?
3. Are platforms like Zabbix, Asana, or Google Data Studio used for visibility?
4. Is metric data linked to decision-making and improvement planning?
5. Are performance goals adjusted based on evolving business needs?

Cybersecurity and Risk Management

1. Do you follow frameworks like NIST or CIS Controls to guide security posture?

2. Are tools like Wazuh, Snort, or SIEM systems part of your defense strategy?
3. Are threat detection and response procedures documented and tested?
4. Have you trained employees on phishing, data handling, and other security topics?
5. Is there a clear escalation process for cyber incidents?

Sustainability and Efficiency

1. Do you use tools like the Green IT Calculator or AWS Sustainability Dashboard?
2. Are energy-saving technologies implemented across systems and facilities?
3. Are cloud-native architectures being adopted where feasible?
4. Do sustainability goals align with IT procurement and operations?
5. Is green IT discussed as part of strategic planning and reporting?

Customer Experience and Digital Enablement

1. Do you actively collect and analyze customer feedback (e.g., NPS, CSAT)?
2. Are IT systems enhancing customer satisfaction through automation and speed?
3. Is infrastructure optimized for smooth omnichannel engagement?
4. Are personalization and accessibility integrated into digital services?
5. Are IT leaders involved in shaping digital customer strategy?

Innovation and Adaptability

1. Are you tracking emerging technologies and evaluating their use cases?

2. Do you have a safe environment for piloting new tools and approaches?
3. Is there a formal process for scaling successful innovations?
4. Are innovation KPIs (e.g., time-to-experiment, idea-to-implementation) defined?
5. Do you promote a culture of curiosity and iterative learning?

Communication and Influence

1. Do you communicate regularly with transparency and strategic intent?
2. Is your messaging tailored to resonate with both technical and business audiences?
3. Do you use storytelling and data to convey impact and urgency?
4. Are you open to feedback and known as a responsive, credible voice?
5. Is communication part of your leadership development for direct reports?

Vendor & Partner Ecosystem Management

1. Do you manage vendors with a structured performance and SLA framework?
2. Are cost, value, and consolidation regularly evaluated?
3. Is vendor risk (e.g., security, dependency, compliance) tracked and mitigated?
4. Are strategic partners involved early in planning and execution phases?
5. Do you use dashboards or tools to monitor multi-vendor landscapes?

Hybrid & Remote Work Leadership

1. Do your workflows support asynchronous collaboration and autonomy?

2. Are tools like Slack, Zoom, and Miro fully integrated into operations?
3. Is leadership adapted for visibility, trust-building, and engagement remotely?
4. Are policies equitable across remote and in-office staff?
5. Is your team culture thriving despite the lack of physical proximity?

Technology Refresh & Lifecycle Planning

1. Do you have a documented roadmap for hardware and software refresh cycles?
2. Are aging systems flagged for replacement or sunset?
3. Is lifecycle management tied to cost planning and risk reduction?
4. Are procurement and IT finance aligned with long-term refresh needs?
5. Are usage, obsolescence, and ROI tracked continuously?

Leadership Legacy and Empowerment

1. Are you mentoring or sponsoring future IT leaders consistently?
2. Is your knowledge and strategy well-documented for continuity?
3. Have you built frameworks that endure beyond your leadership?
4. Do your actions encourage ownership, initiative, and innovation?
5. Are you actively shaping a leadership culture others aspire to emulate?

Your Leadership Reflection

Leadership isn't just about achieving goals; it embraces the journey, the growth, and the impact you leave behind. Taking time to reflect on your leadership journey is essential for understanding where you've been, where you are, and where you want to go. This reflection is your opportunity to assess your strengths, identify areas for growth, and align your actions with your values and vision.

Start by asking yourself what defines your leadership style. Consider which principles and values guide your decisions and daily interactions. Reflect on how these principles show up in practice. Do you empower others, foster collaboration, and lead with purpose? Think about the moments when your leadership was most effective and what made those moments successful.

Next, examine your ability to adapt to change. IT leadership demands agility and resilience. Reflect on how you've responded to challenges: Did you view them as threats or opportunities? Have you been proactive in preparing your organization for future shifts in technology, markets, or regulation?

Think about your relationships with your team and stakeholders. Strong leadership is built on trust and communication. Reflect on how effectively you've built and maintained trust within your team. Are you approachable and transparent? Does your team feel empowered to raise ideas and concerns?

Assess how well you've aligned IT initiatives with broader organizational goals. Reflect on whether your decisions delivered value to the organization. Consider the outcomes of major projects or transformations you've led and whether they achieved their intended impact. Have you used meaningful metrics to measure success and shared these insights with your stakeholders?

Consider your role in fostering innovation and creativity. Reflect on whether you've created an environment that encourages experimentation and supports calculated risk. Did you provide resources and space for your team to explore new ideas and drive meaningful innovation?

Reflect on your approach to mentorship and talent development. Have you taken an active role in nurturing future leaders? Did you create pathways for growth, share your knowledge, and encourage members of your team to step into leadership roles?

Reflect on your commitment to sustainability and social responsibility. Consider whether you've ensured your IT practices align with ethical and environmental standards. How have your decisions contributed to positive societal impact beyond the organization?

Reflect on how data and metrics have informed your leadership decisions. Have you established metrics that matter? Do you review and act on them regularly? How effectively do you use data to shape strategy and lead outcome-driven initiatives?

Reflect on your contributions beyond your immediate organization. Have you influenced vendor ecosystems, industry forums, or community initiatives? How will your leadership be recognized outside your team and company?

Finally, reflect on your legacy. Leadership is about more than immediate results. It encompasses the systems, culture, and inspiration you leave behind. What do you hope your leadership will be remembered for? Have you built foundations that allow your organization and your team to thrive long after your tenure?

Leadership reflection is not just an exercise in self-assessment, it is a commitment to growth. Taking time to evaluate your journey honestly, you refine your approach, deepen your impact, and ensure your leadership continues to evolve. This reflection is not the end; it marks the beginning of the next chapter in your leadership journey.

Leading with Reflection and Growth

Reflect on the journey that shaped you: the values guiding your choices, the lessons learned through change, and the connections built through trust.

Every decision, every setback, and every success contributes to the leader you are becoming.

The Last Word: Shaping Tomorrow with Vision

Leadership is more than a role; it's a responsibility to inspire, to innovate, and to leave the world better than you found it. As an IT leader, you stand at the forefront of transformation, where technology meets possibility and vision becomes reality. The choices you make, the actions you take, and the values you uphold have the power to shape not just your organization, but the future itself.

Shaping tomorrow requires clarity of purpose. It begins with asking, "**What kind of impact do I want to create?**" Whether it's driving digital transformation, ensuring cybersecurity, fostering innovation, or championing sustainability, your vision must guide every decision. This clarity doesn't arrive in a single moment, it's built step by step, choice by choice, as you align your leadership with the legacy you hope to leave.

But vision alone cannot carry you to the finish line. Action transforms vision into reality. It's found in the long hours spent solving complex challenges, the collaboration that fuels creativity, and the perseverance that pushes through setbacks. Leadership demands risk-taking, embracing uncertainty, and adapting as the world evolves. Continuous learning and adaptability become your competitive edge. In these moments, you shape not only outcomes but also the culture and character of your organization.

Leadership is also about people. It's about building relationships anchored in trust, empowering your team to rise to their potential, and listening as much as you lead. The connections you forge, the mentorship you provide, and the belief you instill in others ripple outward, creating a lasting impact that extends far beyond any single initiative.

As you look toward the future, remember leadership is not a solitary journey. Seek inspiration, stay curious, embrace collaboration, and hold yourself accountable. The challenges we face, from technological disruption to environmental sustainability, require collective effort. Measure your impact and remain accountable for progress. Together, as leaders in IT and beyond, we hold the power to drive meaningful change.

The tools, strategies, and frameworks shared in this book are not just resources, they are catalysts for action. Use them boldly, tailor them to your unique context, and let them fuel your ambition. Your ability to combine technical expertise with a clear, purposeful vision and ethical conviction will define your success as a leader.

The future is unwritten, and that is what makes it so full of possibility. Each decision you make and every step you take are opportunities to shape it. Lead with courage, integrity, and the unwavering belief that tomorrow can indeed be better than today. Technology used responsibly can become a force for good.

Now is your moment to make a difference. Embrace your vision, act with purpose, and shape a future that reflects not only business success but also the very best of what leadership can achieve.

Thank You for Reading!

If you found this book insightful, inspiring, or useful, I would greatly appreciate it if you could take a moment to leave a review.

Your feedback not only helps me grow as an author but also helps other readers discover this book and benefit from its insights. Even a few words can make a tremendous difference.

Thank you for your support and for being a part of this journey!

Regards,

Segvan Johnson

www.ingramcontent.com/pod-product-compliance
Lightning Source LLC
LaVergne TN
LVHW051127210926
848482LV00001B/6

* 9 7 9 8 9 9 2 3 9 5 5 1 8 *